Commercial Roofing Systems

John A. Watson

Reston Publishing Company, Inc.
A Prentice-Hall Company
Reston, Virginia

Library of Congress Cataloging in Publication Data

Watson, John A.
   Commercial roofing systems.

   Bibliography: p.
   Includes index.
     1. Roofing. 2. Roofs. I. Title.
TH2431.W327   1984   695   83-24455
ISBN 0-8359-0857-7

Glossary of Roofing Terms by the kind permission of Mr. Maxwell C. Baker and the National Research Council, Canada. Extracted from *Roofs* by Maxwell C. Baker, Multiscience Publications Limited, Montreal, P.Q. 1980.
Note: The inclusion of the glossary from *Roofs* does not in any way imply approval of any statements made in *Commercial Roofing Systems*.

© 1984 by Reston Publishing Company, Inc.
A Prentice-Hall Company
Reston, Virginia 22090

All rights reserved. No part of this book may be
reproduced in any way or by any means without
permission in writing from the publisher.

10 9 8 7 6 5 4 3 2 1

PRINTED IN THE UNITED STATES OF AMERICA

Contents

Introduction, 1

1. **Looking Back, 3:** Old and New Architectural Designs, 4/Built-up Roofing Materials, 6/Pitch, 6/Advertising and Propaganda, 7/Old Roof Decks, 8/Performance of Old Roofs, 9/Early Insulations, 9/Heating Bitumen, 9/Standards for Materials and Roofing Bonds, 10/Early Attempts to Correct Problems, 11

2. **Interior Environment, 13**

3. **Exterior Environment, 17**

4. **Roof Decks and Substrates, 23:** Types of Roof Decks, 24/Boards, 25/Plywood on Wood Joists, 29/Plywood over Tongue and Groove Decking, 31/Steel Decks, 35/Poured Gypsum Concrete, 41/Precast Gypsum, 43/Poured Concrete, 43/Precast Concrete, 47/Lightweight Concretes, 51/Wood Fiber and Cement Slabs, 52/Asbestos Cement Cavity Decks, 52

5. **Recommended Minimum and Maximum Roof Inclines, 55:** Problems With Level Roofs, 55/General Rules on Roof Inclines, 57/

Contents Roof Traffic, 58/Roofing Examples, 59/Asphalt Softening Points and Roof Inclines, 60/Steep Roofs, 60/Danger from Chemicals, 60/Moss, 61/What To Do About Dead Level Roofs, 61

6 **Product Description and Physical Properties, 63:** Asphalt, 63/Coal-Tar Pitch, 65/Organic Felt: Asphalt, 66/Organic Felt: Tar Saturated, 66/Asbestos Roofing Felts: Asphalt and Coal-Tar Saturated, 68/Glass-Fiber Felt, 69/Smooth-surfaced Roll Roofing: Organic and Inorganic, 70/Mineral-surfaced Roll Roofing, 70/Mineral-surfaced, Selvage-edge Roofing, 70

7 **Thermal Insulation, 71:** Roof Insulation Materials, 73/When Is Insulation Needed? 75/Conventional Systems, 76/What Thermal Insulation Does to a Roof, 77/Roof Membrane Cracking and Buckling, 80/Felt Laying Patterns, 83/Research on Roofing Splits, 83/A Frank Statement About Roof Insulation, 86/Reflective Surfacing, 87

8 **Air-Vapor Barriers, 89**

9 **Ventilation of Roof System, 91**

10 **Roof System Specifications, 93:** The State of the Art (1979–1982), 93/Roofing Research, 94/Contributors to Roofing Problems, 96/Reasons for Roofing Problems, 97/Development of Roofing Specifications, 100/Base Sheet (Asphalt Saturated), 101/Roofing Felt and Ply Sheets, 102/Back Nailing Felt, 103/Roofing Bitumen: Asphalt and Coal-Tar Pitch, 105/Asphalt Quantities, 106/Pitch Selection, 107/Pitch Quantities, 107/Surfacing Materials, 107/Types of Roof Decks, 108/Protected Membrane Roofs, 110

11 **Steel and Aluminum Roofing, 119:** Advantages, 119/Roof Decks, 120/Insulation, 120/Vapor Barriers, 120

12 **Drainage Systems, 125**

13 **Metal Flashings, 129:** Basic Rules for Flashings, 129/Thermal Expansion of Metals and Flashing Details, 132

14 **Application Procedures and Workmanship, 151:** Heating Bitumen, 151/Laying Felt, 152/Laying Base Sheets, 153/Stripping Felts, 154/Flood Coat and Gravel, 155/Mopped Surface Coats, 155/Vapor Barriers and Insulation, 155/Mineral Surfaced Cap Sheets, 157

15 **Materials Handling and Storage, 159**

16 **Reroofing Procedures, 163:** The Reroofing Dilemma, 164/Examples, 167/Urethane Foam, 171

17 **Single-Ply Membranes, 175:** Introduction, 175/Manufacturers, 176/Insulation, 178/Contaminants, 187/CGSB General Standards, 188/References, 189

18 **Cold Built-up Roofing, 191:** History, 191/Selection by Geographic Location, 192/Product Description, 192/Roof Decks and Substrates, 193/Roof System Specification and Application, 194/Equipment Required, 194

19 **Fire Resistance and Fire Ratings, 197**

20 **General Maintenance, 199:** General Directions, 200/Summary, 201

21 **Guarantees and Life Expectancy, 203**

22 **Sources of Technical Literature, 207**

23 **A Glossary of Roofing Terms, 209**

24 **Metric Conversion Tables, 225**

Index, 229

Introduction

At one time a roof was considered to be a building component that diverted water to the drainage system. To do this effectively, the roof deck was always sloped to the drains. It is obvious now that most roofs are flat or level and do not drain quickly and sometimes do not drain at all. Materials and systems that once functioned satisfactorily are now unable to do so because of the additional hazards that have been created by modern building designs. Furthermore, as with many other building materials, the quality is not as good. The final result is obvious. Consider the following statement on flat roofs:

> The art of building involves the enclosure of space, which requires a water shedding or watertight system to keep the enclosed space dry. Traditional steeply pitched roofs shed water rapidly, and their coverings of overlapping impervious units perform well. Flat and low pitched roofs tend to hold water, or shed it very slowly, and the covering must be jointless and watertight. The introduction of bituminous roofing materials made possible the trend to flat roofs. Principles of modern architecture also indicate flat roofs as desirable and they are now the accepted form for many types of buildings. Flat roofs, however, have given rise to varied problems in many countries. Some of these problems are related to

Introduction

the introduction of new materials and designs, and to changes in building practice which have taken place in recent years.

The past two decades have seen an increase in the amount of research devoted to bituminous materials, problems of roofing, and more recently to the development of new roof coverings. Most flat roof systems used in Canada involve the sealing of thermal insulation material between a vapor barrier and a roofing membrane. Problems often result from a failure to understand the basic principles of moisture behaviour in such systems. The inadequacy of specifications to define correct materials, the failure of designers to give full attention to all details of the waterproofing system, and the difficulty of obtaining good workmanship, have added to the basic problem, and make it difficult in field studies to determine the exact cause if failure occurs.[1]

This remark was made 20 years ago. A sincere effort has been made in *Commercial Roofing Systems* to examine the various materials used in flat roofs and how these materials should be assembled to avoid early failure. Practices believed to be responsible for less than satisfactory performance and frequent costly replacements are highlighted. The future of built-up roofing is uncertain because of new roofing materials being introduced; however, a reckless adoption of these new materials before we are certain why the old systems did not work would be irresponsible, and might be disastrous.

---

[1] *Building Research News* No. 8 (Ottawa, Canada: Division of Building Research — National Research Council, October 1963).

1
Looking Back

In order to chart a safe course forward in any endeavor, it is often advisable to check back over what has been done, successfully and unsuccessfully. Unless we learn from past mistakes we are condemned to repeat them. This applies to all fields of human endeavor, including building design, construction, roofing, and roofing maintenance.

It has been said, "The wheels of the Gods (and justice) grind exceeding slow." There is no question that this observation also applies to building design. Unfortunately, the "wheels" of the various parts of buildings do not turn at the same speed. A great deal depends on the need for change to improve public safety and health. These two aspects receive the most attention by government regulatory agencies. Such things as structural safety, fire prevention, electrical hazards, heating and ventilating, elevator design, and exterior cladding are accepted as engineering problems, now well covered by standards arrived at by a consensus of opinion, and by trial and error. Even so, failures occasionally occur with disastrous results.

The severity of the problem is illustrated in the following excerpt from "Why are all those buildings collapsing?" by Walter McQuade in *Fortune*, November 19, 1979.

> Since 1979, the payment by insurance companies for building casualty claims has increased from an estimated $32 million a year to

Looking Back

$235 million with hundreds of millions more still under litigation. Over the decade the amount of premiums paid by architects and engineers for liability insurance has risen from about $25 million annually to about $175 million. Malpractice insurance is now second- or third-largest business expense for many architects and engineers, and is inexorably driving up the cost of building. The number of construction cases filed with the American Arbitration Association—most of them involving building failures—quadrupled from 504 in 1967 to 2,042 in 1978; the 1978 figure involved $130 million in claims and counterclaims. . . .

It is also quite clear that there have been an unacceptable number of roofing failures on all types of buildings since 1940 or earlier. The reasons for the failures are many, but may be due principally to changes in construction designs without considering how they might affect the roof. When failures occurred because of an unfamiliar and untested roofing system, new products were introduced instead of paying more attention to scientific research, or even just plain common sense. Also, many roofing problems are due to excessive cost cutting measures when a building is designed and built.

The following report from a small church provides a good example.

The Parish News, 1982

As most of you know only too well by now, St. Mark's is in the process of meeting a rather serious demand for repairs to the roof and walls of the church. Our building, opened for services on Good Friday of 1968, and costing at the time two hundred thousand dollars. The roof has been showing serious leaks for the past ten years or more. Successive building committees have done what they could to repair and remedy such deficiencies as they arose, but obviously always in a kind of "band-aid" style because of the nature of this uniquely designed building. The services of an engineering consultant were obtained and we were informed on good authority that we would have to replace the existing roof, repair the drainage system from the run-off wells on either side of the skylights, replace most of the cedar shakes (on the walls), cover the stained glass windows with translucent fiberglass panels, and repair the interior drywall panels where necessary. The estimated cost is $120 thousand.

In answer to many comments from parishioners, the building committee did check with the original architect regarding our roofing and structural problems. We were informed in a letter, ". . . your roof would only have been guaranteed for two years." They further stated, "The current problems seem to originate with the economics imposed on the builders and ourselves at the time of construction."

Old and New Architectural Designs

Significant changes in U.S. building design began with the work of H.H. Richardson (1838–1886), culminating in the ten-story Wainright building in St. Louis in

1868, and of Louis Henri Sullivan (1856–1924). Sullivan designed the Chicago Carson Pirie Scott store in 1903 and buildings in other cities using a steel skeleton or frame instead of load-bearing brick or stone walls. Later, architects like Mies Van der Rohe (1886–1969), another modernist in the cubist movement, created flat-roofed steel frame buildings in the United States after 1937.

Floors are now steel with a low- or high-density concrete cover, and roof decks steel alone with thermal insulation and a roof covering. Other forms of construction use precast concrete floor and roof slabs or poured gypsum roof decks. In some parts of the country reinforced concrete is used for the exterior walls or frame, and for the floors and roof. The exterior skin of these concrete frame buildings can be almost any type of weatherproofing system or material designed and attached to allow for shrinkage and plastic flow in the concrete. Concrete lift slabs poured flat on the ground and lifted 90° to a vertical position form the exterior walls of single-story buildings. Roofs are laid on steel or wood decking attached to open web steel joists or laminated wood beams and purlins. The roof systems on these buildings are invariably dead level.

Until the turn of the century most, if not all, commercial and industrial structures were built with some form of masonry load-bearing walls and wood framed and sheathed floors and roofs. Thousands of these buildings are still in use, made acceptable today only by automatic fire alarms and water sprinklers. A few all wood structures still exist, usually associated with a forest industry operation. One of the largest of these is located in Cloquet, Minnesota, and was used by the Wood Conversion Company as late as 1960.

Nearly all of these early commercial buildings had wood framed and sheathed roof decks, either as the principal ceiling and roof, or as a framing or loft over a laminated wood mill deck or perhaps a reinforced concrete ceiling. By present-day standards, they were small in area and had a positive slope built into them to drain off water. Outside drainage was probably preferred to inside drains through the middle of the building. No matter how they were built the deck and roof covering did not appear to be affected in any way by the interior environment, except to be warmed by the heat loss through the ceiling structure. One of the principal reasons for this was the absence of firm control of the interior environment to create an imbalance between the interior and the exterior. In other words, the two spaces were in harmony with no great differences or gradients in relative humidity, vapor pressure or air pressure. Heated buildings created a temperature gradient, but leakage through walls, ceilings, and roofs, none containing thermal insulation, did not damage the structure.

No doubt certain buildings such as bakeries, leather tanneries, dairies, pulp and paper mills, and others with wet manufacturing processes, produced condensation of moisture in cold climates, but the materials used in the building enclosure were not as susceptible to damage by moisture as are the more sophisticated materials and complicated assemblies being used today. The mass of the old materials acted as a balance wheel which leveled out the response to changes in the weather.

There is an old roofing felt mill in Portneuf, Quebec, that has brick walls three feet thick. In later years the building housed machines to manufacture both tarred and asphalt felt roofing products for flat and steep roofs. The mean annual minimum outside temperature in the area is recorded at $-30°F$. The mean annual maximum outside temperature is $+90°F$, a temperature spread of 120°. Actual minimum and maximum roof temperatures would be greater than the mean quoted and the temperature of the building surfaces more or less, depending on color and other factors.

Built-up Roofing Materials

Up until the period between 1940 and 1950 when things began to change, built-up roofing (BUR) felts for both tar and asphalt saturation were composed of carefully selected waste paper, new wood fiber, and cotton or wool rags. There were no zippers on old clothing so the buttons were ripped off and the material put through a rag cutter before being further defibered or broken down. The rag contributed to tensile strength and bulk for better saturation of roofing felt. The speed of the saturator and drying-in machine was much slower than it was later when saturating methods changed for a minimum 140% requirement, but allowed the felt to travel through the machine much faster. The 140% saturation meant that if the basis weight of felt was 9 lb per 100 ft^2, the saturant weight was 12.6 lb. Manufacturers would likely aim for the 140% minimum standard saturation when it was distinctly possible they had been achieving greater saturation with the older felt. Minimum standards are apt to lower quality than increase it. For example, the minimum standard for heavier shingle felt was 165%, but it was generally found for a shingle to be stable it required at least 170% or more. This was not hard to obtain with good quality felt.

Most roofing companies dropped rags after 1945 due to the introduction of synthetic fibers for clothing and other uses. These were not suitable for asphalt saturation and there was not enough old cotton and wool available. This reduced the tensile strength of the roofing sheet, particularly in the cross machine direction, and increased its susceptibility to moisture. The increased demand for flat built-up roofing materials during and after World War II increased the proportion of lower quality waste paper, some of which had color-inked surfaces and clay fillers or coatings on smooth stock. The general tendency, perhaps not followed by all manufacturers, was toward greater production at the lowest possible cost and sometimes a lower quality. There is little doubt that the quality of asphalt and roofing felt has diminished since 1939.

Pitch

Coal-tar pitch, once a popular roofing bitumen, has diminished since 1939 so that today there are only a few producers in the United States and none in Canada. Only 5% of BUR in the United States uses coal-tar pitch and tarred felt but many of these roofs may still be in existence.

Some people still believe pitch to be superior to asphalt for waterproofing in BUR. The author's preference for asphalt is based on accepted ASTM standards for both materials, close association with laboratory technicians examining the performance of both by weatherometer exposure and other means, and by checking the application of many roofs and their performance over a 24-year period. Finally, few roofers have as much faith in pitch as they do in asphalt, and there is no doubt about which they would rather work with. Coal-tar, hot or cold, is toxic as are the volatile fumes and cold pitch dust.

There is one curious quality of coal-tar pitch that can be observed on some old buildings with wood decks. The two dry nailed felts over unsaturated paper, plus two or three mopped felts and gravel, was a standard application. The mopped pitch gradually penetrated the two dry felts and the paper down to the board deck where it firmly stuck. If the boards were wide and changed slightly in moisture content and dimension during seasonal or other influences, the membrane would crack at each board. A similar penetration occurred with coal-tar mopped felts over thermal insulation, particularly wood fiber and cork. This movement robs the felt plies of interlayer adhesive and reduces the effectiveness of the membrane.

A four-story warehouse building has been observed with festoons of black cobweb-like material that are actually strands of pitch hanging from the top floor ceiling. The pitch flowed from the roof covering through the cracks between the roof sheathing boards.

Advertising and Propaganda

Coal-tar pitch advertising and propaganda promoted dead flat roofs by claiming pitch was not soluble in water while asphalt was. It was claimed that asphalt could not be used on dead-level decks because the asphalt would dissolve in water. The fact that pitch had been used for underground waterproofing was held as proof that pitch was insoluble.

Under certain conditions water may be absorbed by the bitumen itself or by minute quantities of inorganic salts or fillers in it. The normal solubility of water in bitumen is in the order of 0.001 to 0.01 percent by weight and is so small as to be negligible. The presence of water soluble salts in any quantity will result in a large capacity for water absorption by osmosis. For this reason oil refineries de-salt the crude oil before refining it. Fillers also can absorb certain quantities of water, the amount varying with the composition and granular size of the material. As a result it has been found that bitumens in permanent contact with water absorb it in varying amounts, and various claims as to the relative water absorption properties of coal-tar pitch and asphalt have been made. Results of recent tests on asphalt and coal-tar pitch have reported water absorptions of 0.5 to 2.4 grams/sq ft for commercial coal-tar pitches and 2.0 to 3.9 grams/sq ft for commercial asphalts after one year. This rate of absorption is very low and there is very little difference between the two. It is more significant that the rate of water penetration into the bitumen is also very low.[1]

The proponents of pitch did not seem to mind that smooth-surfaced asphalt roofs were laid on sloping roofs with no gravel cover and exposed to rain, snow, and sun. Pitch on the other hand could only be used safely on roofs under one-half inch per foot slope for fear of the pitch sliding down the slope. A steep roofing pitch was tried but was quickly abandoned because of its brittleness. Pitch could only be used with a protective gravel covering or underground where the sun could not reach it.

---

[1] P.M. Jones, "Bituminous Materials," Canadian Building Digest No. 38, Division of Building Research, National Research Council of Canada, February 1963.

Looking Back

**Table 1.1
Roof Inclines**

Barrett specifications	1955	Later (no date)
Pitch–gravel	0 to 2 in.	0 to 1 in.
Asphalt–gravel	1/2 to 3 in.	0 to 3 in.
Asphalt–smooth (no gravel)	1 to 6 in.	1/2 to 6 in.
Cold process–(asphalt)	—	0 to 8 in.
S.I.S. (17-in. slate)	2 to 9 in.	2 to 9 in.

In 1958 Allied Chemical, Canada, Ltd. published 41 Barrett roofing specifications, including both pitch and asphalt, in a little red book. Barrett later published 109 revised specifications in a larger green book with no date shown. Minimum and maximum inclines for pitch and asphalt are indicated in Table 1.1.
Note the reduction of the maximum slope for pitch from 2 in. to 1 in. per foot, and the reduction of the minimum for asphalt and gravel from $1/2$ in. to 0 in. per foot. The slope for smooth-surfaced asphalt roofs was reduced from 1 in. to $1/2$ in. Cold-process roofs laid with cut-back asphalt cement were allowed on dead-level roofs, but cold-applied S.I.S (seventeen-inch slate) roofs were limited to a minimum of 2 in. per foot. Dead level for cold-applied roofs is dangerous.

With regard to the solubility of asphalt and pitch, the facts show that they are both highly resistant to water in all forms,[2] and because producers knew this they published roof specifications for levels roofs to be competitive with pitch manufacturers. People on both sides of the fence knew a positive slope or an incline to drains made the best roof but their competitive stands allowed dead-level roofs to be built. Building designers were pleased that there were roof specifications for their dead-level roof decks, supported in their minds by a 20-year roofing bond or guarantee. This is an example of a competitive business situation helping to cause a painful and long lasting reduction in the efficiency of roof coverings by promoting an unsuitable structural base. This and the introduction of thermal insulation into the roofing system in the late 1930s were two serious errors.

The claim that cracks, breaks, and leaks in pitch and gravel roofs are self-healing has not been demonstrated to the author on either new or old roofs. The peculiar quality of pitch to flow while cold has usually been to the detriment of a roof membrane rather than to its advantage.

Old Roof Decks

The roof decks and floors in many old brick structures used what was called "mill construction," consisting of rough-sawn two-inch lumber spiked together. Floors were strapped and tongue and groove (T&G) flooring laid over the strapping. Roof decks, however, were left with no board covering and the roof membrane was laid directly on the upper edges of the lumber, generally 2×4 or 2×6, which did not

---

[2] "Asphalt and Allied Substances," 6th ed. Van Nostrand Reinhold Publishing Co., N.Y., 1960.

line up exactly on the top edge. The humps and hollows in the deck were not conducive to good roofing. Occasionally a high relative humidity inside the building created a condensation problem on the underside of the roof membrane due to the temperature drop through the thick wood deck. The later addition of strapping and acoustic tile on the ceiling would make matters worse unless a vapor and air barrier was introduced into the system. If there was no obvious insulation nobody thought about a vapor barrier.

Performance of Old Roofs

Several things made asphalt and pitch roofs last a long time on the old wood decks. First, the roofing felt was generally better quality and more flexible because of its rag content. Second, roofs were laid by hand mopping on relatively small areas. There was no machine laying of felt at high speed. Third, the roof deck always sloped to drain. Fourth, there was no thermal insulation in the system where moisture could be trapped between the roof membrane and a vapor barrier. Fifth, the pace of construction was slower and the critical path method of construction had not been invented, whereby the roofing is scheduled for a particular day, regardless of the weather. When laying a built-up roof the weather is one of the most critical elements in the entire enterprise, and it is the one thing that is not known when a roofing failure or problem is investigated some years later.

Early Insulations

Some old roofs were laid on concrete decks which presented no problem, particularly if they were slightly sloped. Roofing felts were mopped directly to the concrete. Trouble began when thermal insulation was introduced into the system. Materials such as fiber board, cork, and foamglas were first. Attempts were made to build a slope on level decks by adding a cinder concrete (coke breeze) fill, but the sulfur in the cinders destroyed the felt. Later, other lightweight concrete and vermiculite fills were tried but there was generally too much water in the mix that could not be dried out before the roof was laid. Shrinkage in these fills caused cracks in the roofing and the water caused buckles and blisters. Mixtures of vermiculite and hot asphalt were tried with disastrous results. This so-called thermal mix could not be rolled out smooth enough while hot, and when cooled it was not possible to hot mop felt to it because the asphalt melted again and fouled up the mops with asphalt and loose vermiculite. The uneven surface and low resistance to point loading made it unsuitable for roofing.

Heating Bitumen

In the early days asphalt and pitch kettles were fired by burning wood. The rate of heating was slower with less damage to the bitumen; however, temperature control was more difficult and the kettles were known to catch fire, particularly pitch kettles because of the low flash point of pitch (Pitch $-248°F$; Asphalt $-450°F$). It is generally accepted that the first place a hot roof can be ruined is at the kettle. If an inexperienced worker is loading and heating the kettle, all the care in the world on the roof will not correct the worker's mistakes. Some errors can be made or avoided in the use of tankers, but even with these, changes can be made to asphalt without the operator being aware of it. A fall-back in softening point is one result.

Looking Back

Standards for Materials and Roofing Bonds

Standards for roofing materials were considered relatively unimportant in the construction industry until 1941 when the American Society for Testing and Materials (ASTM) adopted standards for felt and bitumen. Standards from the National Bureau of Standards (NBS), the Canadian Standards Association (CSA), and the Canadian Government Specifications Board (CGSB) came later. Performance standards for these materials have yet to be clearly defined. An empirical form of performance standard for 10, 15, and 20 years was developed by roofing materials manufacturers on the basis of their own experience and sales policies, but was not backed with any technical rationale. This was started by the Barrett Company in 1916 and all other manufacturers were obliged to follow suit. If they did not it was construed that they had no confidence in their materials. Roofing bonds were withdrawn in Canada in 1960. Some U.S. manufacturers still supply a form of roofing bond up to 25 years but the bond charge per square is considerably more than in 1916 or 1960, and the penal sum or dollar recovery is very much less than the replacement cost of a roof and insulation. The insulation manufacturers and the deck manufacturers never took part in the bonding system.

The maximum bond period of 20 or 25 years appears to have become accepted or entrenched as a universal standard of performance for a built-up roof, although this period bears no relationship to the probable life of the building. It is ludicrous when present-day roofing costs are about $200 per square and reroofing costs approaching $500.

The original standard that took the form of a roofing bond was probably originally developed to reduce confusion in specifications and create confidence in a new form of roofing for buildings. In the early stages the roofing manufacturer had more or less complete control of the whole business but lost it when roofs and roofers proliferated. Unfortunately, the roofing bond eventually did more to destroy the owner's confidence in built-up roofing than it did to improve the general welfare and respectability of the roofing industry.

There is no question that standards were useful as a guide to manufacturers but of little interest to anyone else. In any event, the quality of the roofing materials themselves had little to do with their assembly in a system combined with many different types of thermal insulation, vapor retardants, and roof decks. The workmanship related to all the dozens of available components was something that was left in the laps of the gods and an occasional construction or roofing foreman.

It has always been true that roofing was something alien to the other building trades and something that was supposed to appear as if by magic at the right time in the construction schedule. Most roofing specifications not prepared by an experienced person very often contained only one line: "All areas indicated on the plans shall be covered with a 20-year tar and gravel roof." Reference to thermal insulation, vapor protection, deck preparation, sheet metal flashings, skylights, and other bits and pieces might be found elsewhere in the architect's specifications. Material standards for specific products rarely appeared. It was often apparent that the specifications writer had no knowledge of the materials, how they were put together, or how they could be repaired or replaced sometime in the future.

Unfortunately, the cost of construction resulted in a roof system that was incapable of lasting for many years. Dead-level decks are one of the most disagreeable legacies from past economies and will be with us for a long time. They are extremely difficult to correct. There are other problems.

Early Attempts to Correct Problems

The following is an extract from *National Roofer*, December 1955.

> The Federal Construction Council in a recent technical report has requested the National Roofing Contractor's Association to: "accept responsibility for correlating the major roof problems and for directing a coordinated program of research and investigation."
>
> The Federal Council's report and recommendations followed three industry conferences held, at the invitation of the Building Research Advisory Board, in Washington, last Spring. These conferences made history in that for the first time in the history of the roofing industry, representatives of all interests connected with the design, manufacture and application of roof decks and roofing were invited to discuss problems peculiar to some or common to all and to suggest ideas for the solution of problems or indicate paths to solutions.
>
> At the three industry conferences the attendance included representatives of the American Institute of Architects, Associated General Contractors, the Portland Cement Association, National Roofing Contractors' Association, Insulation Board Institute, Copper and Brass Research Association, Vermiculite Institute, in addition to individuals representing manufacturers of asphalt and tarred roofing products, steel, aluminum, perlite, cork, and other materials and representatives of building owners — industrial firms such as Du Pont, General Electric and the American Telephone Co.
>
> The Federal Construction Council's Task Group turned a critical eye on every aspect of the built-up roofing sandwich — on the design of the deck and on each component of the completed deck and covering. The Council believed that there were four major causes of built-up roofing failure:
>
> 1. Blistering of roofing caused by entrapment of air and water vapor in insulation.
> 2. Damage to roofs caused by foot traffic in servicing air conditioning and other equipment on roof tops.
> 3. Cracks on roof decks and roof coverings caused by lack of provision for expansion and contraction movement.
> 4. Water damage to dead flat roofs during and after installation due to lack of provision for drainage.
>
> However, it was pointed out, there were other causes of roof failure: Faulty roofs had been traced to poor workmanship, improper

Looking Back

materials, improper application, inadequate flashing, omission of expansion joints and infiltration of the roof with vapor from the interior of the building.

Other recommendations were made by the Council in the above report but regrettably there is little evidence that much was accomplished. This task is now being duplicated by the Roofing Industry Educational Institute (6851 S. Holly Circle, Suite 250, Englewood, CO 80112, Richard L. Fricklas, Director).

2

Interior Environment

As a selective separator of dissimilar environments, a roof system is subjected to variations of almost all the environmental factors; the differences from one side to the other determine the duties of the roof and the properties it must possess. The environmental differences of greater importance to the durability of a roof system relate to rain penetration, heat flow, vapor flow, air flow, radiation, and fire, all of which involve actual or potential flows of mass or energy.

In all cases flow is from the higher to lower potential, and the net result is a tendency to equalization of potentials. In complex constructions such as roofs, the performance of any one material influences the environment and the performance of all the other materials in the system.

The environment in which any roof must serve is determined by the environments being separated, the properties of all the materials in the system, their relative positions, and the behavior of the roof structural system. By judicious selection and arrangement of materials, the roof designer can greatly ease the requirements of the various elements and thus broaden the choice of materials and methods of construction to gain a durable system.

Most materials offer resistance to flow, and gradients of potential occur across materials or constructions interposed between dissimilar environments. These gradients include air pressure, vapor pressure, temperature, and thermal bridges. When

Interior Environment

the temperature in a building differs from the outside, pressure differences occur between inside and outside as a result of differences in the density of the air. This is called chimney or stack effect, since it is the same mechanism that causes a draft in a chimney. With inside temperature higher than outside, chimney effect produces a negative inside pressure relative to outside and infiltration at lower levels, with a positive pressure and exfiltration at higher levels. The opposite occurs with inside temperature lower than that outside. This is confirmed by observation of actual buildings, where severe condensation may occur between panes of windows at upper levels although not at lower levels. Exfiltration of air above the neutral zone (one-half to two-thirds the height of a building) is the source of moisture. Similarly, condensation between panes is more excessive on leeward than on windward sides of buildings. Negative pressures caused by wind will also cause air to move or flow into parts of roof systems when the resistance to such flow is poor.

Condensation problems associated with air leakage in heated buildings will be most prevalent in upper floors, especially on leeward sides, and will increase with severity and duration of winter weather and with increasing building relative humidity.

Pressures inside buildings and air leakage patterns are affected by any imbalance of the air supplied and exhausted by air-handling systems. These systems are sometimes designed and operated to provide an excess of supply air, and thus to pressurize the building and reduce infiltration, particularly that resulting from stack effect at lower levels of multistory buildings during cold weather. The pressurization that results from a given excess of supply air will depend on the tightness of the building enclosure. Under these conditions it is imperative that an air-vapor barrier below thermal insulation in a roof system be impervious to air. The perfection that is required is seldom achieved.

Water is one of the several gaseous constituents of air, the other principal ones being nitrogen, oxygen, and carbon dioxide. In the normal range of atmospheric temperatures and pressures, water can exist in three differnt states: gas, liquid, and solid. The maximum amount of water that can exist in the gaseous state (vapor) is limited by the temperature. Thus if any air-vapor mixture is cooled, a temperature will be reached at which it will be saturated, and if cooling is continued below this point, water will condense. If the temperature at which the air becomes saturated (i.e., the dew point) is above the freezing point, the vapor will condense to a liquid; if it is below freezing, it will condense as ice in the form of hoar frost.

The ratio between the weight of water vapor actually present in the air and the weight it can contain when saturated at the same temperature is called the relative humidity of the air. It is usually expressed as percentage. As the vapor pressures are set by the quantities of vapor in the air, the relative humidity is also given by the ratio between the actual vapor pressure and the saturation vapor pressure at the same temperature. Thus, if the temperature and relative humidity are known, the actual vapor pressure can be calculated from the product of the relative humidity (expressed as a decimal) and the saturation pressure. These saturation vapor pressures and the corresponding quantities of water in the air are given in psychrometric tables published by the U.S. Department of Commerce Weather Bureau. They also appear in the *Guide and Data Book* of the American Society of Heating, Refrigerating, and Air-Conditioning Engineers in the form of a psychrometric chart. The few examples taken at random (Table 2.1) show the vapor pressures for various

Table 2.1
Vapor Pressure — inches of mercury and pounds per sq. ft.

Dry bulb temp.	Relative humidity (%)														
	80		70		60		50		40		30		20	
	in. Hg	lb sq ft	in. Hg	lb sq ft	in. Hg	lb sq ft	in. Hg	lb sq ft	in. Hg	lb sq ft	in. Hg	lb sq ft	in. Hg	lb sq ft
90	1.091	76.91	.989	69.72	.866	61.05	.707	49.8	.575	39.54	.432	30.45	.277	19.53
80	.833	59.08	.707	49.8	.616	43.43	.517	36.45	.417	29.40	.310	21.85	.211	14.87
70	.595	41.95	.517	36.45	.448	31.59	.373	26.30	.287	20.23	.228	16.07	.150	10.57
60	.417	29.40	.360	25.38	.310	21.85	.256	18.05	.211	14.87	.157	11.07	.108	7.614
50	.287	20.23	.256	18.05	.219	15.44	.180	12.69	.143	10.08	.113	7.97	.0735	5.18
40	.195	13.75	.180	12.69	.150	10.57	.130	9.16	.103	7.26	.073	5.18	.054	3.82
30	.130	9.16	.108	7.61	.103	7.26	.085	6.00	.066	4.69	.049	3.46		

Interior Environment

Table 2.2
Dew-Point Temperatures Fahrenheit

Relative Humidity	Dry Bulb Temperature					
	50	60	70	80	90	100
100%	50	60	70	80	90	100.
90%	47.2	57.1	67	76.9	86.8	96.73
80%	44.13	53.9	63.68	73.5	83.3	93.11
70%	40.7	50.32	60.0	69.64	79.3	89.04
60%	36.76	46.3	55.74	65.27	74.8	84.38
50%	32.26	41.6	50.83	60.19	69.56	79.0
40%	27.41	35.85	45	54.09	63.26	72.48
30%	21.33	29.16	37.57	46.44	55.33	64.3

conditions of temperature and relative humidity. The exterior pressure must be subtracted from the interior pressure to arrive at the net pressure being exerted on the building envelope. This can be added to the air pressure, which is measured in inches of water, since air and vapor pressures operate independently of each other. The dew point temperatures shown in Table 2.2 show the narrow margin of safety when the relative humidity is high.

It is the function of a building envelope when acting in conjunction with heating and ventilating equipment to maintain a more or less uniform internal environment regardless of weather conditions. While the inside temperature may be regulated for human comfort within a 10° range of 65° to 75°F, the outside ambient temperature can swing annually from $-40°$ to $+100°F$. Surface temperatures of roofs may reach 200°F or more depending on color, heat-storage capacity of substrates, and other factors. It is obvious that the thermal insulation selected to modify these potential extremes must be extremely efficient, and must be placed at a location in the system where it will not be degraded over a long period of time, and where it will complement and protect, rather than destroy, adjacent components. The economic justification for a particular insulation material in a roof system design must be supported by concrete evidence that its efficiency will not deteriorate in service.

3

Exterior Environment

In the early part of the century, commercial and industrial buildings were much smaller in floor area than they are in the 1980s. The ceiling and roof-supporting systems were either separated, or the combination ceiling and roof covered a space that was not as narrowly controlled, atmospherically. The effect of the interior environment on the roof system was therefore negligible. Today, the separation of the controlled interior and the uncontrolled exterior environments is often accomplished in the space of 2 in. or 3 in. of solid materials. The forces explained in Table 2.1 can be considerable, and must be added to the forces exerted by the weather. Walls, on the other hand, are thicker because they are often load bearing. The need for thermal insulation in addition to the structural elements is not as critical, and if there happens to be a transfer of heat and vapor, the effect on the exterior cladding is not as catastrophic as it is on a roof membrane, because it is not a flexible, moisture-sensitive material, and it does not hold water like a flat tray.

Weathering of walls naturally takes place, but they do not face the same hazards as does a roof covering owing to the difference in solar orientation. The hazards that vary according to geographic location include the following:

Sunshine The annual hours of sun vary from 1,800 hours in Seattle to 3,000 hours in Miami. The solar altitude at noon on June 22 in Seattle is approx-

Exterior Environment

imately 66°. In Miami it is 90°. For solar altitude and azimuth for various degrees of latitude refer to the Smithsonian Meteorological Tables. Table 170 provides a series of charts for each 5° of latitude (except 5°, 15°, 75°, and 85°), giving the altitude and azimuth of the sun as a function of the true solar time and the declination of the sun.

The greater number of hours and the higher altitude of the sun create a greater potential for high roof surface temperatures, temporary softening of roofing bitumen, and more rapid oxidation of all roofing materials. Low-boiling-point constituents are dissipated in the form of gas until both bitumen and felt become hard, brittle, and low in tensile strength. Other more complicated chemical changes take place as a result of ultraviolet radiation. The heat generated by direct exposure to solar radiation and that reflected from walls adjacent to the roof contribute to the vaporization of liquid moisture in the system, resulting in deformation of soft bitumens and separation of felt plies. Surface temperatures can exceed the softening point of asphalt and pitch, causing sliding.

Diurnal changes in surface temperatures of a roof system are greater in summer than the ambient temperatures because of the absorption and storage of heat during the day and rapid radiation to a clear night sky. Such thermal cycling is bad for any roof. Constant changes in temperature expand and contract metal flashings. The movement is sufficient to pull nails out of solid wood, concrete blocks, and brick. Once the flashing is loose a good wind will rip it off.

Wind Wind may reduce surface temperatures on a summer day, which is an advantage, but wind chill cooling in winter without a snow cover will increase the rate of heat loss. Constant wind, even at moderate speeds, will scour a roof, removing the gravel cover at roof edges, building corners, and at penthouses. The combination of solar heating and wind will materially damage smooth-surfaced asphalt roofs and will tear pieces off mineral-surfaced selvage roofing, even when they are hot mopped.

If flashings are not well designed and fastened, a strong wind will remove them. If the roof base is nailed only to the substrate, the loss of roof-edge flashings may trigger the loss of the roof membrane by suction.

The effect of wind on buildings and roofs is an extremely complicated subject that cannot be covered adequately in this more or less elementary coverage of roof design and construction; therefore, the reader is referred to the following sources for information and guidance.

The American National Standards Institute (ANSI)

The National Building Code — U.S.A.

The Uniform Building Code

The Southern Standard Building Code

The Basic Building Code (BOCA)

Factory Mutual Engineering Corporation

Underwriters Laboratories (UL)

The National Building Code — Canada

The National Research Council — Division of Building Research — Canada

Water Roofing materials are manufactured and roof systems are designed to shed water. They are not designed to hold water like a tank for some useful storage purpose. Roof water is always dispatched as waste water except in very unusual circumstances. A built-up roof made with asphalt or pitch would not make a suitable collector for water.

When water is allowed to stand on a flat roof, one must suffer these consequences:

1. In the event of a minor defect in construction or an accidental puncture, there is a reservoir of water to flow into the system. It may be stored in the insulation if an effective vapor barrier exists and the insulation is capable of holding water (most are), or it may flow into the building interior some distance from where it entered the roof system.

2. The combination of water and ultraviolet light can have a more damaging effect on most materials than either one alone.

3. The slow evaporation of ponded water can leave concentrated solutions of chemical pollutants from the atmosphere or from the building itself. On a sloped roof such pollutants are washed away with each rainfall. If necessary, water sprinkling can be installed to provide cooling as well as cleaning.

4. Ponded water is often responsible for the degradation of sheet-metal flashings and makes each flashed projection through the roof a potential leak.

5. Each inch of water adds 5.2 lb per square foot (25.18 kg per square meter) to the live load on the roof structure. Roofs have been discovered with up to 6 in. of water (31.20 lb) because of blocked drains. Some of these were discovered because the roof was leaking over the flashings. Others were found accidentally, and when the roof was cleared the force of the water damaged the drainage system.

6. The deeper the water becomes the greater the deflection is in the deck, which allows more water, and so on.

7. Water on a roof encourages the growth of vegetation. Grass and small trees have been found with roots growing through the roof membrane into wood-fiber insulation.

8. It is difficult to repair leaks in ponded roofs unless the water can be removed by pumping, syphoning, or by vacuum equipment, and the roof thoroughly dried. It is sometimes time consuming to locate a leak in a graveled roof when it is covered with water.

9. Water on a roof in hot, dry areas may be considered to have some value as a coolant, a fire-protection device, and a buffer against wind damage, but its presence is usually accidental and cannot be relied on for such services. As a general rule, the disadvantages far outweigh the advantages.

10. In cold climates roof water turns to ice when temperatures stay well below freezing. The formation of ice on the surface does no particular harm to a roof, but after it is covered with an insulating blanket of snow the heat loss from the building through an insulated system is enough to melt the ice next to the roof. The net result can be a layer of water, a layer of ice floating on the water, and a blanket of snow. Even at subzero air temperatures, the temperature gradient is only 35° to 45°F (22° to 27°C). Under these conditions the roof and flashings would have to be perfect in every way.

Exterior Environment

Exterior Environment

11. Hailstones can cause severe damage to roofs in the north central states and the Canadian Prairies. A graveled roof on a smooth, solid base would have the best chance and would be hard to puncture, even by the largest hailstones. Another solution is the protected membrane system with a heavy gravel or concrete block ballast. Soft insulation under the membrane should be avoided if the roof is to be exposed to possible hailstorms.

Conclusions

By simple experience and observation or by studying the voluminous and detailed weather reports available for every section of the country, the probability of certain weather phenomena can be determined. For the building designer, such a study will be rewarding, because it will undoubtedly indicate that there may be better ways to protect a building from the elements by using different building shapes and roofing materials much better able to withstand water, ice, hail, snow, wind, high and low temperatures, air pollutants, and solar radiation than are offered by layers of soft roofing felt and asphalt or pitch exposed to the weather.

All roofs are a combination of many parts or components. Each one on its own can be a good product, but when combined with other components in the wrong way, in the presence of heat and moisture gradients, it frequently falls short of its potential for weatherproofing a building.

One cannot dismiss a roof covering, whether it be flat or steep, as something that will "just happen." Its composition must be considered from the time the first line is drawn at the drafting table. It is certain that when the Parthenon was designed, so cleverly, the Greek architect knew how to prevent water from entering his masterpiece. Like the Parthenon in Athens or the Taj Mahal near Agra, what good is fabulous symmetry and beauty if the roof leaks?

An architect, engineer, or home builder must master many trades before a successful structure can be put together. There is a strong suspicion that in recent years the art of roofing has, on this continent, slipped to a low level of priority as far as designers are concerned. It is true that in the last 50 years buildings have changed drastically in shape, size, and construction, and we are demanding greater control of the interior environment for the sake of our own comfort but at no greater cost.

To achieve this goal new materials and construction techniques are being constantly introduced, creating unforeseen problems along the way. It seems that as soon as we solve one problem, two more pop up. Any text on building construction, or roofing, like this one, is partially obsolete when it comes off the press.

This modest and elementary volume attempts to show what traps to avoid and emphasizes the need for all students of construction and of roofing to spend time studying the interaction of one material with another, and above all to remember that the roof may be overhead and out of sight, but the slightest error in design or application will bring disaster very quickly. There is no question that better informed design and application personnel will produce better roofs and fewer costly errors. The roof that costs the least is the one that requires very little attention and no repairs for a long time.

Weather Data Sources

For weather data refer to:
1. Key to Meteriological Records Documentation No. 4.11, Selective Guide to Climatic Data Sources, U.S. Department of Commerce, Environmental

Science Services Administration. (For sale by the Superintendent of Documents, U.S. Government Printing Office, Washington, D.C. 20402)
2. Selected Climatic Maps of the United States, Environmental Data Service, National Climatic Center, Asheville, N.C. 28801
3. Climatic Information for Building Design in Canada, Supplement No. 1 to the National Building Code of Canada, N.R.C. No. 13986

4

Roof Decks and Substrates

Like the foundations of a building, the roof deck is the foundation of or for the roof covering, and is an important element in its performance. A good deck should possess the following qualities, or at least as many as possible.

1. Contain as little moisture as possible at the time that the roof is laid: 10% to 15% maximum or as close as possible to what will be expected in service.
2. Have minimum dimension change when curing or drying and during subsequent changes in the interior environment.
3. Provide good nailing or a smooth even surface for hot mopping, cold adhesives, or liquid coating.
4. Provide for fastening wood nailers where required for roofing or sheet metal.
5. Have maximum resistance to vertical air flow through the deck.
6. Provide strength characteristics to allow for complete replacement or resurfacing of old roofing without damaging the deck or making it necessary to redeck the building.
7. Not be permanently damaged by rain or snow or a leaking roof.
8. Possess a dry, dust-free surface with good shear strength. That is, it must not

Roof Decks and Substrates

lift off the substrate because of wind uplift or condensation of moisture vapor at the membrane-deck interface.

9. Possess the capability to absorb and dispel small amounts of moisture without deterioration, steel reinforcement included.
10. Be fire resistant and have a minimum fire spread rate.
11. Have predictable strain movements resulting from uniform loads, point loads, rolling loads, wind loads, sudden impact, and plastic flow.
12. Be resistant to decay by microorganisms or oxidation.
13. All metal fastenings of the roof deck to the supporting structure must be adequate in number, size, and length, be of rust-proofed steel or a suitable metal alloy, and be designed to hold the deck in position with minimum movement laterally and vertically.
14. Have adequate flexural strength to keep deflection between supports to $1/360$ of the span, *or less*. Permissible loads in engineering tables can be expected to be exceeded on most roofs, which are usually designed for a deflection of $1/240$ of the span.
15. Be able to resist damage during transit, handling, and installation and common roof traffic.
16. Have economical installation.
17. **Have a minimum incline to drains, 2%.**

A building designer has a special responsibility in selecting a roof deck and designing the supporting structure to properly carry the roof system. It is not enough to design a structural system and deck to satisfy engineering requirements, fire regulations, economic considerations, and esthetics, without also selecting an appropriate roof membrane, thermal insulation, and drainage system, and considering the effect of the exterior environment on the whole.

It does not occur to most designers that flat roofs often require major repairs or replacement during the life of the building. Some require regular maintenance in the form of coatings. A designer must therefore look ahead 20 or 30 years to the time when such work must be done.

Types of Roof Decks

1. Boards on wood joists or purlins. Square edge, shiplapped, or tongue and groove (T&G).
2. Plywood on wood joists or purlins.
3. Plywood over T&G decking.
4. Steel, with plywood, gypsum board, or insulating board overlay.
5. Poured gypsum.
6. Precast gypsum.
7. Poured concrete.
8. Precast concrete, with insulation or concrete fill over.
9. Cellular concrete.

10. Lightweight concrete.
11. Vermiculite concrete.
12. Perlite concrete.
13. Wood fiber and cement slabs.
14. Asbestos cement cavity decks, with insulation or concrete fill over.
15. Thermal insulation on structural deck.

Table 4.1 lists the various decks by number in the left column. Columns A to M show what materials, fastenings, and special treatments are required for each type of deck so that a quick comparison can be made. This should help in preparing roofing specifications.

Boards

Wood boards of various species were the first material used as a base for flat built-up roofing. When the roof is applied directly to the boards, they should have a minimum thickness of 3/4 in. (15.24 cm) and a maximum width of 6 in. (15.24 cm), and be dried to the moisture content expected in service (10% to 15%). Tongue and groove jointing is preferred to square edge or shiplap owing to more solid support, reduced vertical air flow, and a superior surface for roofing. Two nails are required at each support. Galvanized or annular ring nails are recommended, long enough to penetrate the framing 1 in. Long smooth shank nails driven into green lumber may withdraw or pop as the lumber dries. This can puncture the roof membrane.

When a bulk-type insulation is located below the deck, any space between the insulation and the deck should be ventilated to the outside. To assist or improve the ventilation, the roof boards should be nailed to strapping on top of the joists. Diagonal sheathing permits the roofing material to be laid parallel to the roof edge but at a 45° angle to the sheathing. **Roofing felt should never be laid parallel to roof boards.**

If the framing space is completely filled with insulation (all forms), the ceiling and spaces over the partitions must be airtight. Air and vaportight ceilings are advisable for all flat-roofed buildings.

Important Considerations

1. Excessive tangential shrinkage from green to oven dry can tear or split vapor barriers and roof membranes nailed to wide roof boards (more than 6 in.). Softwood shrinkage 4.5% to 9%. Hardwood shrinkage 5.4% to 9.3%.[1]
2. No materials of any kind should be hot mopped to wood boards with pitch or asphalt, or adhered with cold adhesives. A double-coated (coated both sides) asbestos or glass base sheet or separator sheet is first nailed to the deck through nailing discs. Wood fiber felt is not particularly suitable for this purpose because it can absorb moisture from the wood or from the building interior, and buckle. The buckles will not disappear. Filled asphalt coatings on roll roofing sheets are not completely vaporproof.

[1] Kennedy, E.I. 1965. *Strength and Related Properties of Woods Grown in Canada.* PAO Canadian Woods — Their Properties and Uses.

Table 4.1 (Part One)

Deck Type No.	A Primer Asphalt	B Primer Pitch	C Dry Sheet	D Coated Base Sheet Organic	E Coated Base Sheet Glass	F Coated Base Sheet Asbestos	G Combination Sheet*
1.	No	No	Yes	Yes	Yes	Yes	Yes
2.	Yes	No	Yes	Yes	Yes	Yes	Yes
3.	Yes	No	Yes	Yes	Yes	Yes	Yes
4.	No	No	No	No	Yes	Yes	No
5.	No	No	No	Yes	Yes	Yes	No
6.	No	No	No	Yes	Yes	Yes	No
7.	Yes	Yes	No	Yes	Yes	Yes	No
8.	Yes	No	No	No	No	No	No
9.	No	No	No	Yes	Yes	Yes	No
10.	Yes	Yes	No	Yes	Yes	Yes	No
11.	No	No	No	Yes	Yes	Yes	No
12.	No	No	No	Yes	Yes	Yes	No
13.	No	No	No	Yes	Yes	Yes	Yes
14.	Yes	No	No	Yes	Yes	Yes	No

*Glass and kraft paper laminated with asphalt but not coated.

Table 4.1 (Part Two)

Deck Type No.	H Self-locking Nails and Caps	I Straight Nails and Caps	J Mop to Deck Asphalt	K Mop to Deck Pitch	L Venting: Base Sheet Deck Interface	M Substrate Cover for Deck
1.	No	Yes	No	No	No	No
2.	No	Yes	Yes	No	No	No
3.	No	Yes	Yes	No	No	No
4.	Yes	No	No	No	No	Yes
5.	Yes	No	No	No	Yes	No
6.	Yes	No	No	No	No	No
7.	No	No	Yes	Yes	No	No
8.	No	No	Yes	No	No	Yes
9.	Yes	No	No	Yes	No	No
10.	Yes	No	Yes	Yes	Yes	No
11.	Yes	No	No	No	Yes	No
12.	Yes	No	No	No	Yes	No
13.	Yes	No	No	No	No	No
14.	No	No	Yes	No	No	Yes

Notes:

2A Plywood joints should be T&G or taped to prevent drip through of primer or hot bitumen.

2C Required only if base sheet or felt is nailed to deck.

3C Required between plywood and T&G deck if felt or base sheet is mopped to primed plywood.

4AB No primer on steel, but may be required on plywood or gypsum board cover.

4EF Fire-istant slip sheet between steel and covering material. PVC is also acceptable.

4H Covering material must be mechanically fastened to steel decks in all specifications.

6DEF Coated base sheet may be required to serve as a vapor barrier for an insulation overlay if the thermal insulation resistance R 2 must be increased.

6H Use self-locking nails and caps for securing base sheet rather than a hot mop of asphalt or pitch.

9L Venting is suggested because of moisture in the deck, but success cannot be guaranteed. (See Section 12.16.) This also applies to 5L, 10L, 11L and 12L.

10AB Priming assumes that concrete is dry and will not foam when mopped with hot bitumen.

13M Insulation cover required if R 1.06 is not sufficient on heated building.

14J Type 3 or 4 asphalt or other non-heat-sensitive adhesive.

14M Rigid insulation covering.

Roof Decks and Substrates

3. Fasten boards to framing with hot-dipped galvanized steel nails or ring shank nails or screws to prevent nail-pop. Dry framing lumber is required because it is the shrinkage of green or unseasoned lumber that causes the nails to pop. The lumber shrinks away from the nail head. Nail to dry strapping to simplify nailing procedures.
4. All end jointing should be made on solid bearing and well nailed.
5. Good ventilation of insulated spaces below the roof deck is required to prevent wood decay and collapse of the deck. An airtight ceiling is essential.
6. Wood joists supporting a roof deck should be lapped, not butted, over beams or supporting walls, or be carried over more than two supports. A simple span with only two common supports may result in roof cracking due to ordinary deflection and end rotation of the joists.
7. Wood decks with nailed roof membranes can be readily stripped and reroofed without damage to the deck (see Figure 4.1). However, the roofs are not

Figure 4.1 Roof decks like these with odd angles and elevations and severe shrinkage of lumber do not help a roof covering. The old roof being shoveled off in little bits is pitch and gravel on tarred felt. Heavy-duty vacuum equipment should be used to remove all loose gravel and dust before the membrane is disturbed.

immune to wind damage, especially if the base sheet is nailed without 1 in. to 1½ in. discs. Staples should never be used.

Plywood on Wood Joists

8. Insulating with bulk wool between the joists is much cheaper and more efficient than rigid insulation under the membrane, provided the space can be well ventilated, outside to outside, and an air/vapor barrier plus fire resistant ceiling cover. Enough bulk insulation can generally be installed to provide up to three times the thermal resistance of a reasonable and practical thickness of rigid insulation board. It is also more efficient in preventing heat gain in summer.

Plywood on Wood Joists

Plywood should have not less than five plies of veneer and four glue lines. The plywood must be waterproof exterior grade. An unsanded surface is acceptable. The plywood thickness for rafter spacings should be based on the American Plywood Association recommendations for plywood subflooring rather than for plywood roof decking, unless the plywood is first covered with an air/vapor barrier

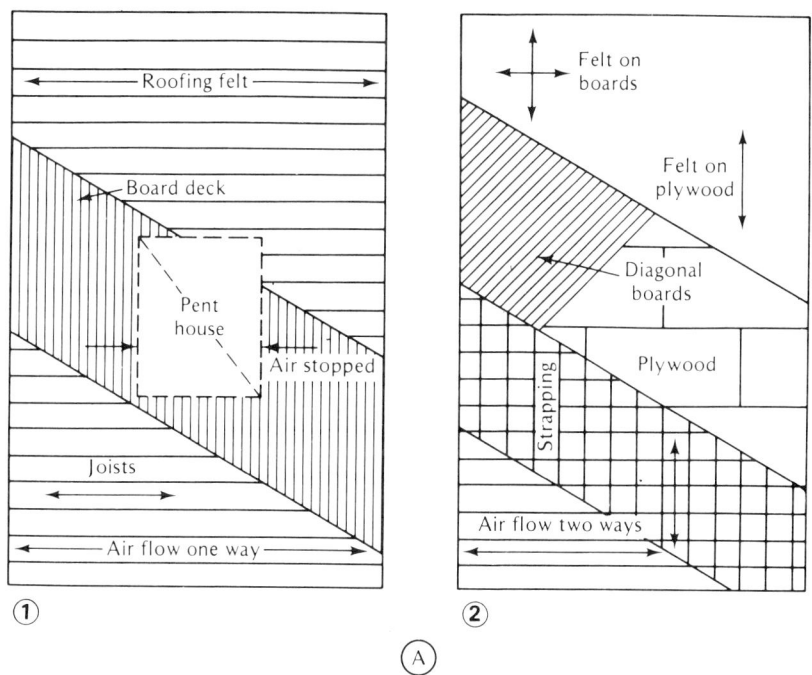

Figure 4.2 Improvement in air flow under flat roof when decking is laid on strapping over joists: (A) Roof plans. In Plan 1, vents are only effective on two sides. In Plan 2, vents are effective on all four sides. (B) Simplified detail. In Plan 1, air circulation can be blocked by penthouses, vents, skylights, and solid bridging. (C) Improved air flow. (D) Old school roof deck shows unnecessary changes in direction of roof boards. Slight tilt at roof edge (left) caused pitch and gravel roof to slide off, exposing gravel stop. Shrinkage of boards is excessive for good roof performance. (E) Excessive shrinkage and distortion at end joints of 2- by 6-in. T&G decking. This section had been covered with fiberboard insulation so shrinkage did not affect the roof, which failed due to ordinary exposure.

Roof Decks and Substrates

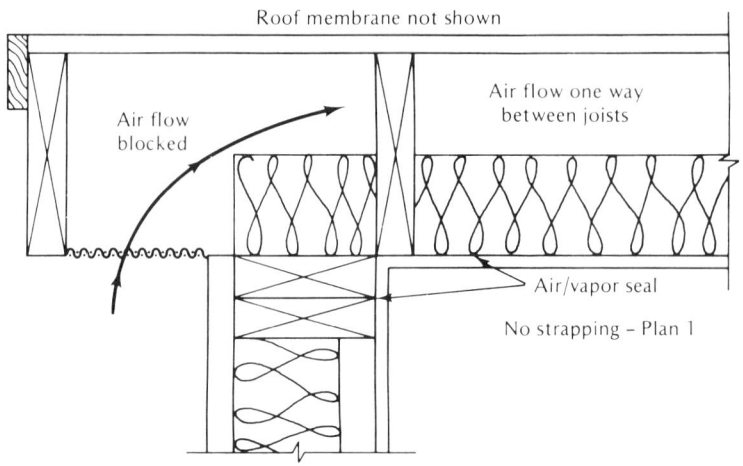

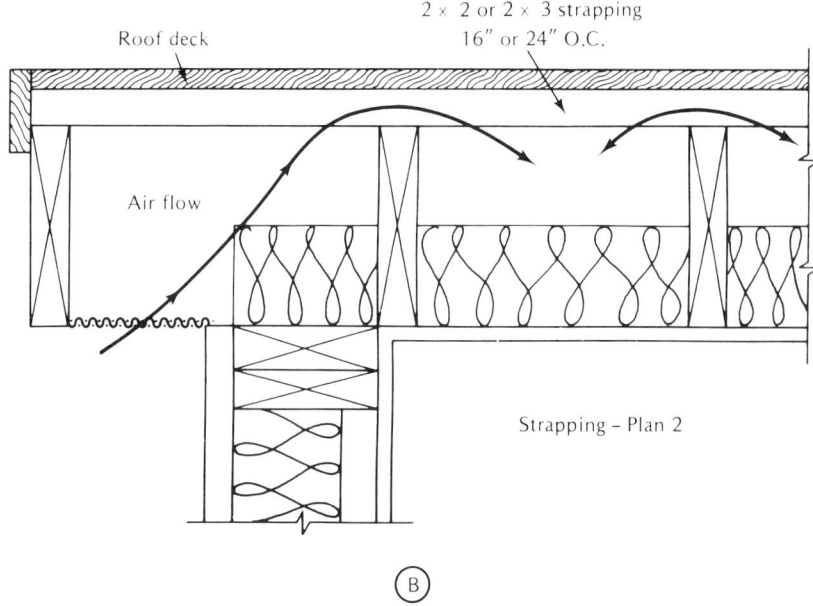

Figure 4.2 (cont.)

and thermal insulation. The "glued floor system" is recommended for flexible roof coverings. (Refer to *Plywood Residential Construction Guide*, APA-Y405 677.)

Since plywood panels are recommended to be spaced $1/16$ in. at the ends and $1/8$ in. at the sides, square edges should be taped with fiberglass tape to prevent an asphalt priming coat from penetrating the joints. Alternatively, use T&G jointed sheets or start with building paper and a nailed base sheet.

Because plywood has a very smooth surface the felts can be laid with constant thickness of bitumen adhesive, which provides improved bonding between plies and a roof with more constant thickness and strength.

Plywood over Tongue and Groove Decking

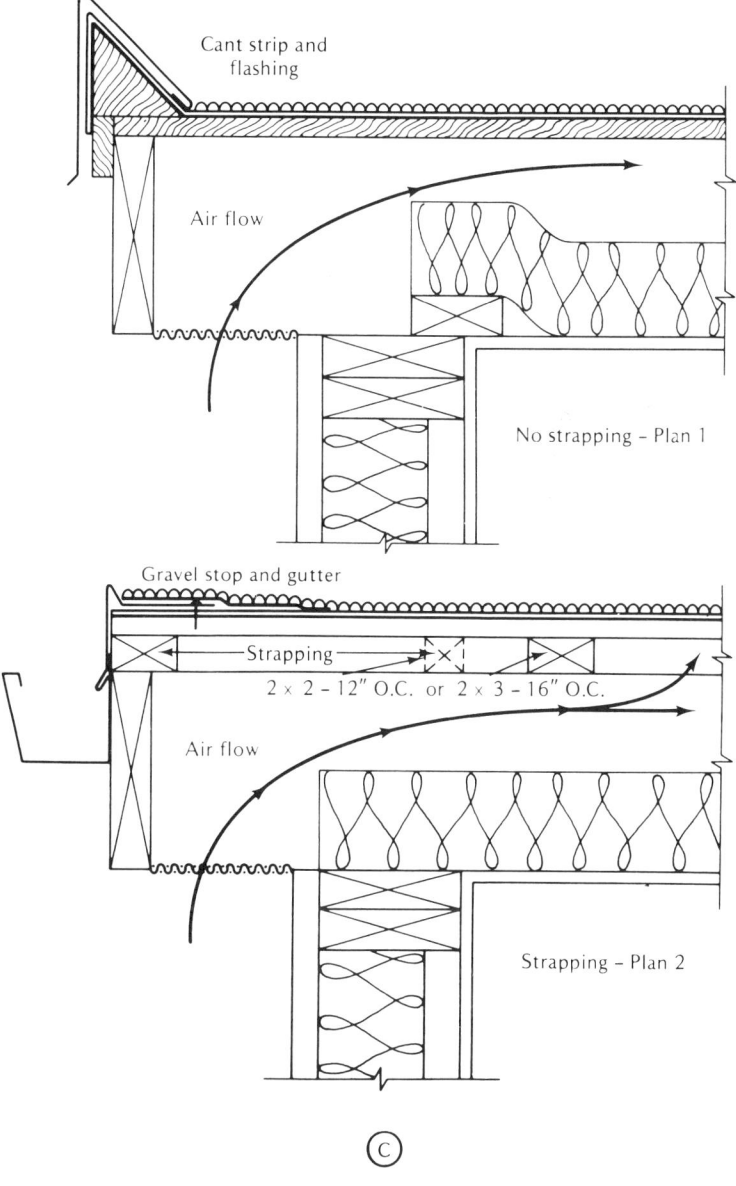

Figure 4.2 (cont.)

Plywood over Tongue and Groove Decking

The use of a plywood cover on kiln-dried T&G decking permits a lower grade of lumber to be used and offers the advantage of a mopped-on roof membrane for maximum wind resistance. A satisfactory roof can be built with three or four plies of No. 15 asphalt felt without a base sheet. Use 3/8 in. unsanded exterior plywood nailed or stapled to a T&G deck over unsaturated building paper, with the long

(D)

(E)

Figure 4.2 (cont.)

dimension at 90° to the T&G boards. Run the felt 90° to the long dimension of the plywood sheets (see Figure 4.3).

Plywood over Tongue and Groove Decking

The face and edges of the plywood should be primed with cut-back asphalt primer as soon as the plywood is laid. Nail or staple the plywood 4 in. on center on the ends, 6 in. on center on the sides, and 12 in. on center throughout the balance of the sheet. The panels must be drawn up tight against the T&G with galvanized barbed nails, ring shank nails, or power-driven staples.

A plywood cover on T&G decking serves to tie the boards together, spreading the live loads between boards that sometimes shrink out of the grooves. Reroofing or

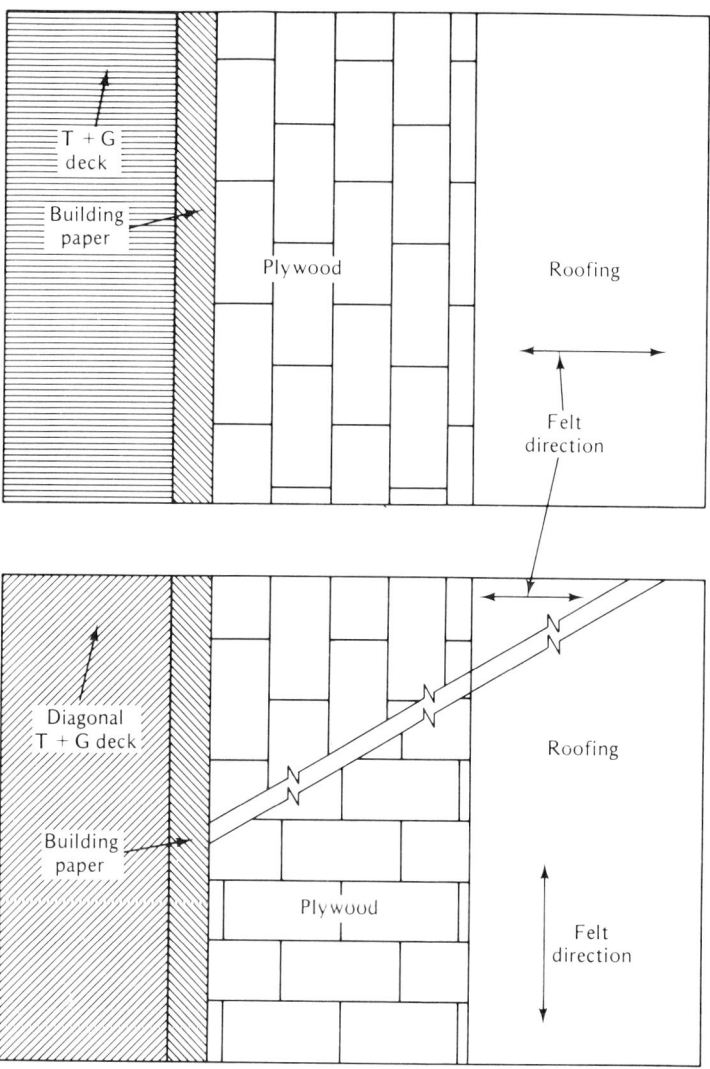

Figure 4.3 Combination of T&G decking and plywood makes application of roofing more flexible.

Roof Decks and Substrates

resurfacing an old roof is made much easier. It also makes an excellent base for a protected membrane system.

Important Considerations

1. Of 13 deck types, No. 2 and 3, both plywood, are the only ones where a roof covering can be mopped or nailed to the deck.
2. The thickness of plywood on joists should not be reduced below standards for flooring (see APA-E30). A roof is not a floor but occasionally carries heavier loading and is subjected to extremes of weather that do not affect the floor. The common belief that any old thing will carry a roof needs correcting. Many people believe they are still hanging wood shakes on pole rafters.
3. One of the main advantages of plywood is its smooth surface, which makes it ideal for obtaining a constant thickness of hot mopping between the felt plies. This insures a roof covering of even thickness and constant tensile strength, plus an absence of voids that even when filled with dry air can expand into blisters and buckles.
4. The smooth dry surface is ideal for single-ply elastomeric membranes, loose laid, adhesive tacked, or fully adhered.
5. T&G wood decks $1^{1}/_{2}$ in. and $2^{1}/_{2}$ in. thick of kiln-dried wood, not wider than 6 in., make a good base for a roof membrane composed of a nailed base sheet (asbestos or glass) and two or three plies of No. 15 felt. Asphalt-saturated and coated base sheets must be followed by asphalt felt, not tarred felt.
6. It is important that T&G decks do not contain more than 10% or 15% moisture when laid and covered, to avoid undesirable shrinkage. Excessive shrinkage causes the boards to carry loads individually rather than together. This can split the roof. All end jointing should be made on solid bearing or be tongued and grooved or splined. Simple spans with end joints on the same support must be avoided to prevent the end rotation thick boards splitting the roof membrane. This precaution also applies to all forms of precast concrete, asbestos cement, and steel decks.
7. T&G wood decks can be covered with unsaturated paper, $^{3}/_{8}$ in. unsanded plywood, primed with cut-back asphalt, and a mopped felt roof membrane.
8. Thick wood decks can be insulated, provided the thermal resistance of the wood is not more than one third of the total resistance of the system. If it is much more than one third, there is a chance of condensation at the membrane.
9. Wood is useful in that it is capable of absorbing water vapor and evaporating it without producing instant condensation. It should never be painted because this could trap moisture and cause decay. The development of microorganisms in wood is an involved process and deserves careful study. Briefly, there are four major requirements for growth.
    a. A food source — cellulose, lignin, or other wood constituents.
    b. A moisture content (dry weight basis), generally between 25% and 100%.

Roof Decks and Substrates

brittle at low temperatures to make a good adhesive in this situation. Coal-tar pitch is not recommended. These materials are also hazardous in case of internal fires.

6. Cold adhesives such as chlorinated rubber, even with high adhesive strength, cannot be applied over a sufficient area and thickness to guarantee the attachment of vapor barriers or insulation under extreme wind suctions. The combination of deck flutter due to wind gusts and the aging or interrupted curing of the adhesive is believed to be partially responsible for wind damage to roofs on steel decks. A total dead load for a steel deck, vapor barrier, insulation, and four-ply graveled roof rarely exceeds 10 lb per ft^2. Negative wind pressures under steady or gust conditions can easily exceed this figure, accounting for fluttering or vibrations of the deck. This condition does not exist in any other roof system. Solutions lie in much heavier and stiffer steel construction and a ballasted roof system. Roof decks lack the concrete ballast fill in steel floor systems that dampen vibrations from foot traffic. Steel decks also lack the beneficial effect of a high degree of transverse load distributing ability and of the advantages of higher dead load-to-design live load ratios.

7. Mechanical fastening of insulation to steel decks is now recommended by many authorities as a more reliable method than using an adhesive. Unfortunately, this brings forth other undesirable conditions.

   a. Any air/vapor barrier is punctured by not less than 25 screw-type fasteners per square. The hexagonal shaped, and one of the largest discs made, has an area of 7.31 in.^{2}. The manufacturer recommends one fastener for each 4 ft^2 of insulation, which means that one square of roof (100 ft^2) is held down with only 1.23 ft^2 of restraining device (see Figure 4-6). Other special nail and disc fasteners for steel decks have approved Factory Mutual perimeter nailing patterns. (See FM Loss Prevention Data 1-28 dated March 1975.) Owing to the nature of the insulation and its direction in relation to the direction of the flutes in the steel deck, the number and type of fasteners are changed. Typical areas of fasteners for various types of insulation materials are as follows (per 100 ft^2) (see Table 4.2).

   b. One manufacturer states, "Embed roof insulation in solid moppings of hot steep asphalt on the bearing surfaces of the decking." On steel decks they recommend mechanical fastening of all insulation panels as additional security against possible wind uplift and lateral transfer (movement) of roof insulation. Inclines of 1 in. or greater require some provision for mechanical securement of the insulation.

   c. Some fasteners are the screw thread-type and some are an annular ring friction-type. Heavy-gauge metal decks may require predrilling. Both types must draw the steel deck and the insulation tightly together no matter how badly it is deflected, twisted, or dented. The nailing disc must not compress the insulation to the point where either the disc or the nail will puncture the roof membrane when a load is applied.

   d. Another type of fastener has a separate 2$^{1}/_{8}$-in.-diameter steel disc, curved

c. A temperature, generally between 50° and 95°F.

d. An adequate supply of oxygen.

10. When the underside or ceiling of a T&G deck must be kept clear, it is tempting to place electric conduit on the top side and attempt to hide or bury it in the insulation. This must be avoided because it always leads to problems.

Steel Decks

A steel roof deck is a structural platform that must be covered with a substrate capable of bridging the open flutes in the deck and providing both thermal insulation and firm support for the roof membrane. This is a difficult requirement for any low-density insulation when approximately 42% of the material has no support. This is a common configuration used on many flat roofs and is reported to be covering approximately 65% of new industrial buildings.

Steel decks are designed for expected uniform live loads, the actual stress in the section, or when deflection at midspan is the criterion. The design deflection of $1/240$ of the span is permitted except when a suspended ceiling is hung directly from the deck. The deflection is then limited to $1/360$ of the span. It is recommended that the roofing contractor be required to plank the decking before hoisting the roofing materials in order to avoid denting the top surface of the decking. Specific directions are given for puddle welding the deck to the steel framing and for crimping the side flanges to each other in order to achieve a shear diaphragm. Some side joints are simple laps that are fastened with sheet-metal screws. Heating and ventilating buildings under construction during the winter months are specifically mentioned by the manufacturer in order to prevent the condensation of moisture on the underside of the steel roof deck, which, it is stated, results in a chemical reaction with the zinc coating, presenting either an unsightly appearance or inhibiting the application of a paint finish. Nothing is said about the possibility of condensation of construction phase moisture in the insulation, which is a very real hazard in cold climates.

The application of engineering data in the design and use of steel roof decks appears to be on the optimistic side in view of the problems that arise. A few of these reported by roofers and observed by the author are as follows:

1. There is often too much flexibility in the deck when it is designed for a deflection ratio of 240 under a uniform load.

2. Designing for uniform loads does not properly reflect concentrated on-site loads of roofing materials and application equipment, nor does it consider the rolling loads that tend to twist the ribs and flutes of light steel decks.

3. Steel deck manufacturer's fastening directions must be religiously followed or movement between adjacent steel sheets can disturb the insulation substrate and fracture the membrane.

4. If the top flanges are not flat and in the same plane, no adhesive will secure the insulation substrate sufficiently to prevent a blow-off.

5. Hot asphalt or pitch are not reliable adhesives on narrow ribbed steel with absorptive insulation materials. Low-melt-point (type 1 and 2) grades may run through the deck at improperly nested end laps. Types 3 and 4 asphalts are too

Table 4.2

	No. per Sheet	No. per Square	Area (ft²)
Expanded glass 2 by 4 ft	6	75	2.09
Wood fiber 2 by 4 ft	4	50	0.25
Celo-Therm; Permalite; Fesco Board 2 by 4 ft	4	50	1.38
Glass fiber; Perlite; Composite Fesco Foam; Permalite PK; Millox 3 by 4 ft	6	50	1.42
Composite fiberglass/ Urethane 3 ft by 4 in.	5	42	1.17
Glass fiber 4 by 8 ft	15	47	1.31

shank, and automatic clips for insulation up to $3^5/_8$ in. thick. This is not suitable for light-gauge springy decks. Under ideal conditions, it depresses with any compression of the insulation and reengages when the load is removed. The clips engage the underside of the deck.

e. To be properly supported on the flutes, the insulation should always be run across the flutes so that the end joints may not fall on solid bearing but the longer side will. However, to provide solid support for the laps of the vapor barrier, it should run the opposite way. The industry literature invariably shows the insulation running parallel with the flutes (see Figure 4.4).

8. Much of the foregoing information or discussion admittedly gives the impression that steel roof decks have problems. These are more likely to become the roofer's problems, because the engineering design of the structure and the roof deck will generally be based on acceptable engineering principles. The roofer's problems may be considerably reduced if a dense structural material such as plywood (when permitted by fire regulations) or gypsum board is mechanically secured to the steel to provide a continuous nailable and moppable surface on which to lay a vapor barrier, insulation, or a roof membrane. It is the only way that a ballasted protected membrane system can be constructed on a corrugated steel supporting platform.

Important Considerations

It could be unfortunate that fluted steel is the most common material used for roof decks on commercial buildings on this continent. The preference for steel is due to the absence of moisture, construction speed, and economy. The health and welfare of the roof system is largely ignored by steel deck manufacturers who, like manufacturers of other building materials, prefer not to worry about how a roof covering is constructed and maintained. In order to achieve even further economies, engineering firms whittle the steel thickness down to the minimum allowed by local codes. As stated at the

Roof Decks and Substrates

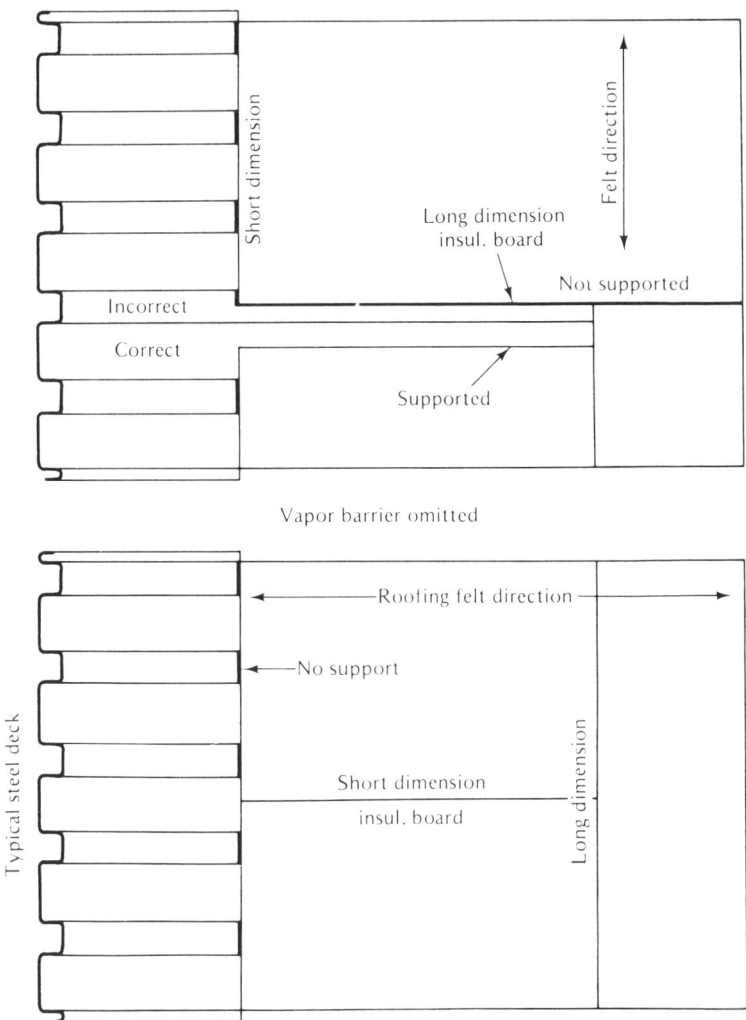

Figure 4.4 Insulation on steel decks.

beginning of this chapter a roof depends on a good foundation or base. No architect or engineer would dare to design a building foundation that could not support the building with a considerable margin of safety.

Even if the steel deck is designed to the highest possible standards it is still only half a deck and must be covered with another material such as rigid insulation, gypsum board, or plywood. When fastened with adhesive alone these substrates are likely to be lifted off by wind, because any rigid sheet material like the three mentioned or expanded or extruded polystyrene, foamglas, or polyurethane, are not flexible enough to conform to the irregularities of the steel deck. The upper surfaces of the flutes seldom lie in the same plane and are easily moved up and down by point or moving loads. For this reason adhesives alone are not acceptable and various forms

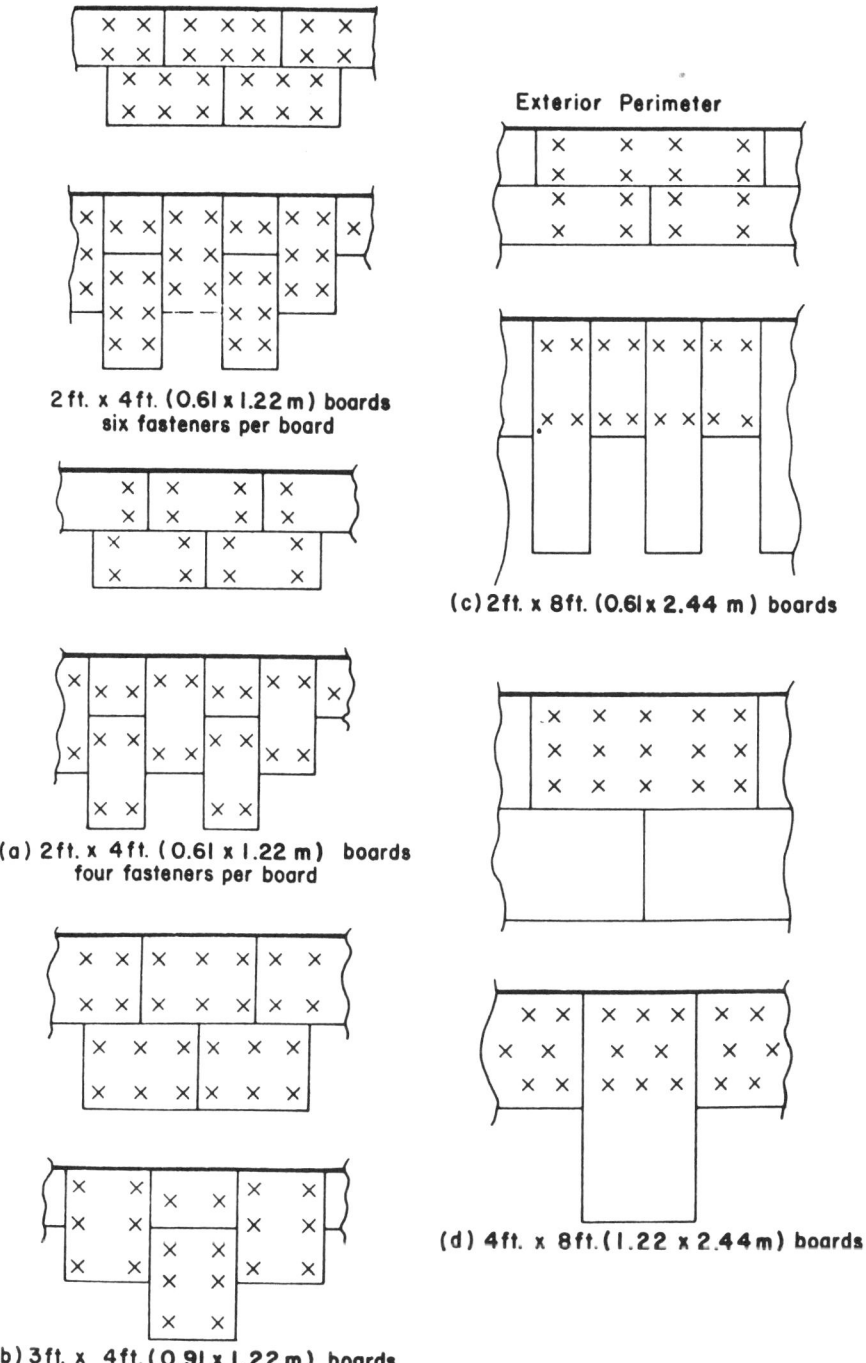

Figure 4.5 Fastening of various sizes of insulation board at exterior roof perimeter. (For mechanical fasteners that are approved for use with the board.)

Roof Decks and Substrates

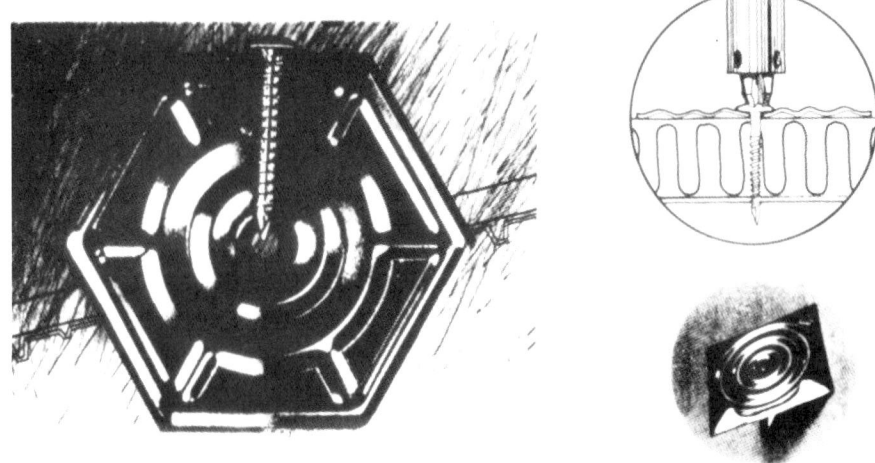

Figure 4.6 One type of fastening device for insulation on steel decks.

of mechanical fasteners must be used, all of which puncture an air/vapor barrier at the steel/rigid board interface in 25 to 50 places in each square of roof. These perforations of the deck and vapor barrier are potential air and vapor leaks into a conventional roofing system. This does not happen in a protected system because there is no place for the vapor to go. Unfortunately, steel decks are invariably dead level to achieve a wind diaphragm and are not entirely suitable for this system, because of the lack of positive drainage.

If a steel deck is covered with a closed cell insulation such as foamglas or extruded polystyrene without a vapor barrier on a heated building, the system will probably fail because interior moisture vapor can pass through the deck at the laps and the end joints and then through the joints in the insulation to the roof membrane where it will condense in cold weather into free water. There are no moisture-absorbing materials in the system to take up moisture quickly and hold it. This quality is present in all other roof systems in varying degrees. The lack of any moisture absorption is particularly dangerous during winter construction when large amounts of moisture are present in many building materials and processes. Condensed moisture vapor will drip back into the steel flutes and can damage the deck and weld fastenings by rust. If the insulation is moisture sensitive it can collapse and fall into the flutes. Wood fiberboard is one material that has done this.

If the insulation is laid in two layers with staggered joints the same thing will happen because the sheets of insulation are not thermoplastic and will not conform to each other to form a tight solid mass through which air will not flow. If an air/vapor barrier is introduced and riddled with mechanical fasteners, the system still allows interior moisture vapor to pass through to the roof covering. This assumes an air and vapor pressure gradient and an outside temperature below the dew point of the inside air.

If an open cell or fibrous glass-type of insulation is used that is capable of holding water, it too will deteriorate and lose its thermal efficiency and eventually produce destructive deformation in the roof covering, whatever it may be.

Roof repairs on steel decks are not easy to make; therefore a low initial building cost may be misleading and in the long run very expensive.

Mechanical and other methods of fastening insulation and rigid sheet material to steel decks deserve special attention. Reference should be made to current Factory Mutual and Underwriters Laboratories specifications and requirements for fastening thermal insulation to steel decks to resist loss by wind suction and fire damage. Underwriters commonly assign Class A, B, or C compliance classifications for individual or combinations of materials. It is important that an owner understand perfectly what these approvals or ratings mean since they have a bearing on insurance rates. The actual performance or efficiency of the components in a roof system, such as special adhesives, fasteners, insulating materials, and roof membranes acting together may not be clear with regard to their waterproofing abilities.

It would appear the ability of the various systems to effectively keep the building dry is not within the scope of Underwriters or Factory Mutual.[2] Each year Factory Mutual prints a Construction Bulletin 1-28 through its Loss Prevention Data.

Poured Gypsum Concrete

Gypsum roof decks are constructed by mixing powdered gypsum rock (calcium sulfate dihydrate) with water. Full details of the chemical process are eliminated for simplicity. The final product for roof decks has a compressive strength of about 500 lb per square inch (minimum) and a density of about 50 lb per cubic foot. The slurry of water and gypsum is pumped to the roof where it is supported on form boards resting on steel structural members, subpurlins, and reinforcing mesh. The subpurlins can be bulb tees, truss tees, or cold-rolled tees (see Figure 4-7). These are generally spaced 32 or 24 in. apart to support the form boards of gypsum, mineral fiber, glass fiber, or cement asbestos. As soon as the form boards and wire reinforcing are in place, the gypsum slurry is pumped into place to a minimum depth of 2 in. Heat is generated in the mix as hydration of the gypsum begins, and after screeding it sets in 12 to 18 minutes.

Curing or hydration as in portland cement mixes is not required; therefore, the covering can be laid almost immediately. The deck-laying crew and the roofing crew must work in harmony with each other to produce a desirable result. Gypsum concrete expands slightly (1%) as it sets, and gypsum concrete decks require expansion joints as with other slab-type roof decks.

Gypsum decks are not primed or mopped with hot bitumen. A nailed, coated base sheet is used as a base for the mopped asphalt felts (see Figure 4.8). For more complete information on poured gypsum roof decks, reference is suggested to manufacturer's engineering data.

---

[2]Underwriters Laboratories Inc., Northbrook, Ill.; Underwriters Laboratories of Canada, 7 Crouse Road, Scarborough, Ontario, Canada M1R 3A9; Factory Mutual Engineering Corporation, Norwood, Mass. 02062.

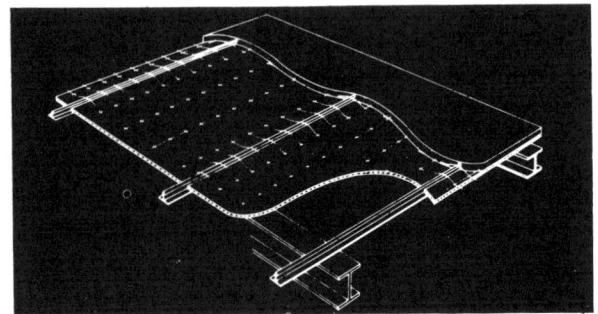

Figure 4.7   Preparing deck for poured gypsum.

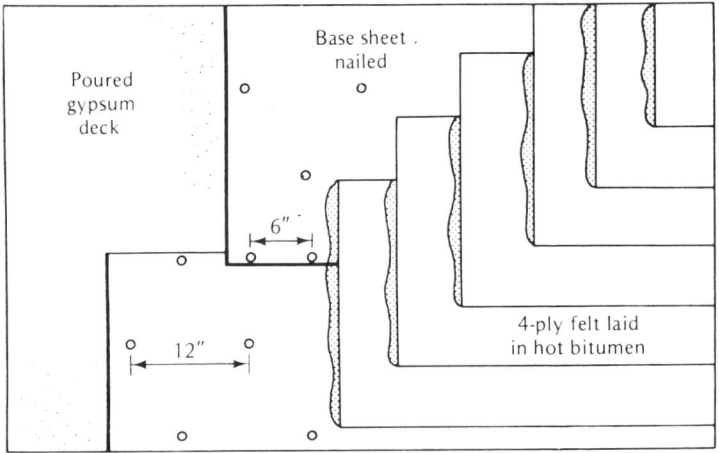

Figure 4.8   Nailing base sheet to poured gypsum.

Precast Gypsum

Precast gypsum planks are made 15 in. wide and up to 10 ft long (38.10 by 304.80 cm). Thickness is 2 in. (5.08 cm). Planks are reinforced with 16-gauge galvanized wire mat and galvanized steel T&G edges on ends and sides. Weight is 11.0 lb per square foot. Maximum span over three supports (two spans) is 7 ft. Deflection ratio is 240. Planks can be clipped, welded, or nailed to supports. The end joints need not be made directly over the supports.

The thermal resistance R of 2-in. plank plus built-up roofing is 2.0. The metal edging will act as a thermal bridge between the interior and the underside of the roofing, which could cause moisture condensation under certain conditions of temperature and humidity. Additional insulation is indicated either above or below the deck to achieve today's standards of thermal resistance in heated buildings. Careful design for individual circumstances is essential to avoid condensation in the system. Manufacturers of gypsum plank roof decks should be consulted.

Metal-bound gypsum plank provides a good dry deck for built-up roofing where a coated sheet is first nailed to the deck with self-locking nails and discs. Felts should be run across the planks. Nailing into the metal edging must be avoided (see Figure 4.9).

Poured Concrete[3]

For the roofer, a reinforced concrete roof deck has many advantages over other types of decks and few disadvantages. Poured concrete permits a wider selection of roofing specifications, with fewer failures if properly designed, applied, and maintained.

Assuming that the water, cement, fine and coarse aggregates, and admixtures have been properly proportioned, mixed, poured in place, screeded, and troweled, the roofer should only be concerned with the water content in the slab after curing has taken place. It is noted in the Portland Cement Association bulletin that air-entrained concrete with a 3- to 4-in. slump contains 265 to 340 lb of water per cubic yard of concrete. On a 100-ft^2 basis (one roofing square), a 6-in. slab contains 491 to 630 lb of water when poured in place. A non-air-entrained concrete contains 300 to 385 lb of water per cubic yard or 556 to 713 lb per 100 ft^2 of 6-in. slab. Additional water may be sprayed on the surface in the curing process in hot weather which concerns the hydration of cement and water.

The design strength of concrete is reached in 28 days under normal conditions, and provided the temperature remains above 40°F (4.44°C) and there is water present. The compressive strength continues to increase under moist curing.

Since a roof may be applied within the 28-day period, water may migrate to the surface where it can be changed to vapor under solar heat. In addition to the simple formation of blisters, the presence of moisture at the concrete-roofing interface may interfere with the bond to the concrete. If the primer won't stick, the concrete is

---

[3] So that the designer, builder, roofer, and roofing inspector are properly informed on the complexities of concrete and how its physical properties might affect a roof covering, it is suggested that they study the information in the engineering bulletin, *Design and Control of Concrete Mixes*, published by the Portland Cement Association, Old Orchard Road, Skokie, Ill. 60076.

Roof Decks and Substrates

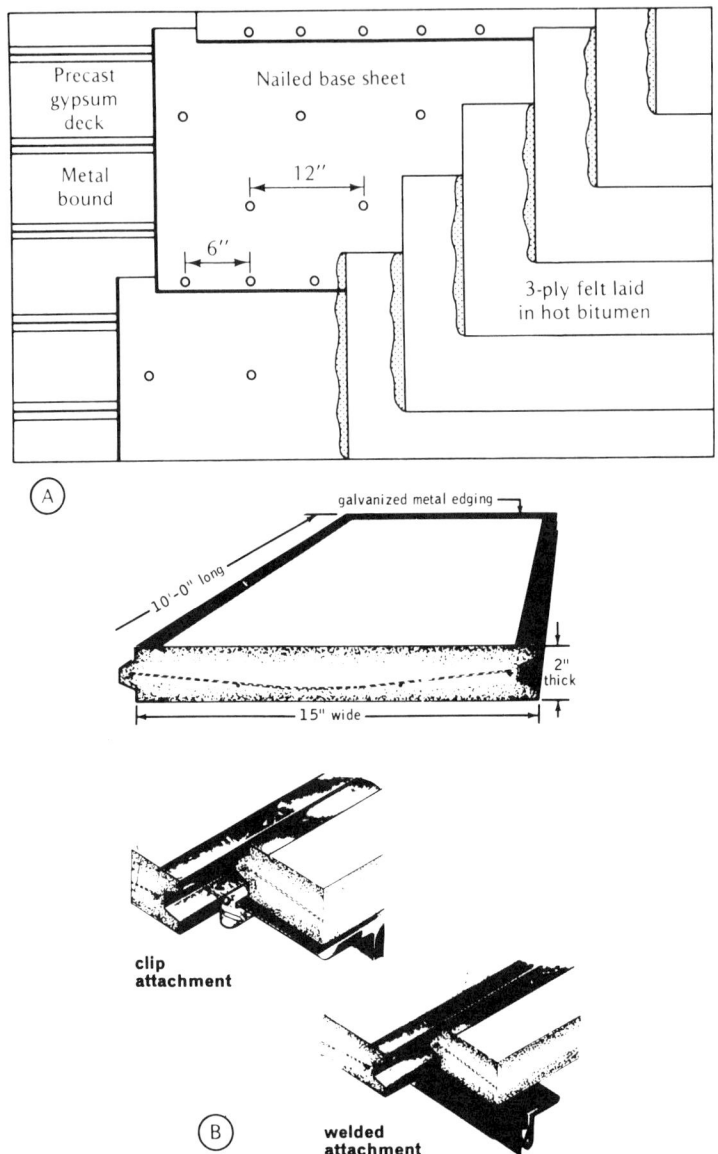

Figure 4.9 (A) Nailing base sheet to precast gypsum. (B) Precast gypsum deck and attachment.

too wet. Priming also assists in reducing problems of adhesion due to the presence of dust caused by premature floating and troweling.

The advantages of poured concrete roof decks are

1. High ratio of dead loads (deck, insulation, and roofing) to design live loads; roughly 1:1. This eliminates flutter due to wind and vibration from roofing and maintenance activity.

architectural details

roof decks
USG Metal Edge Gypsum Plank C 1650

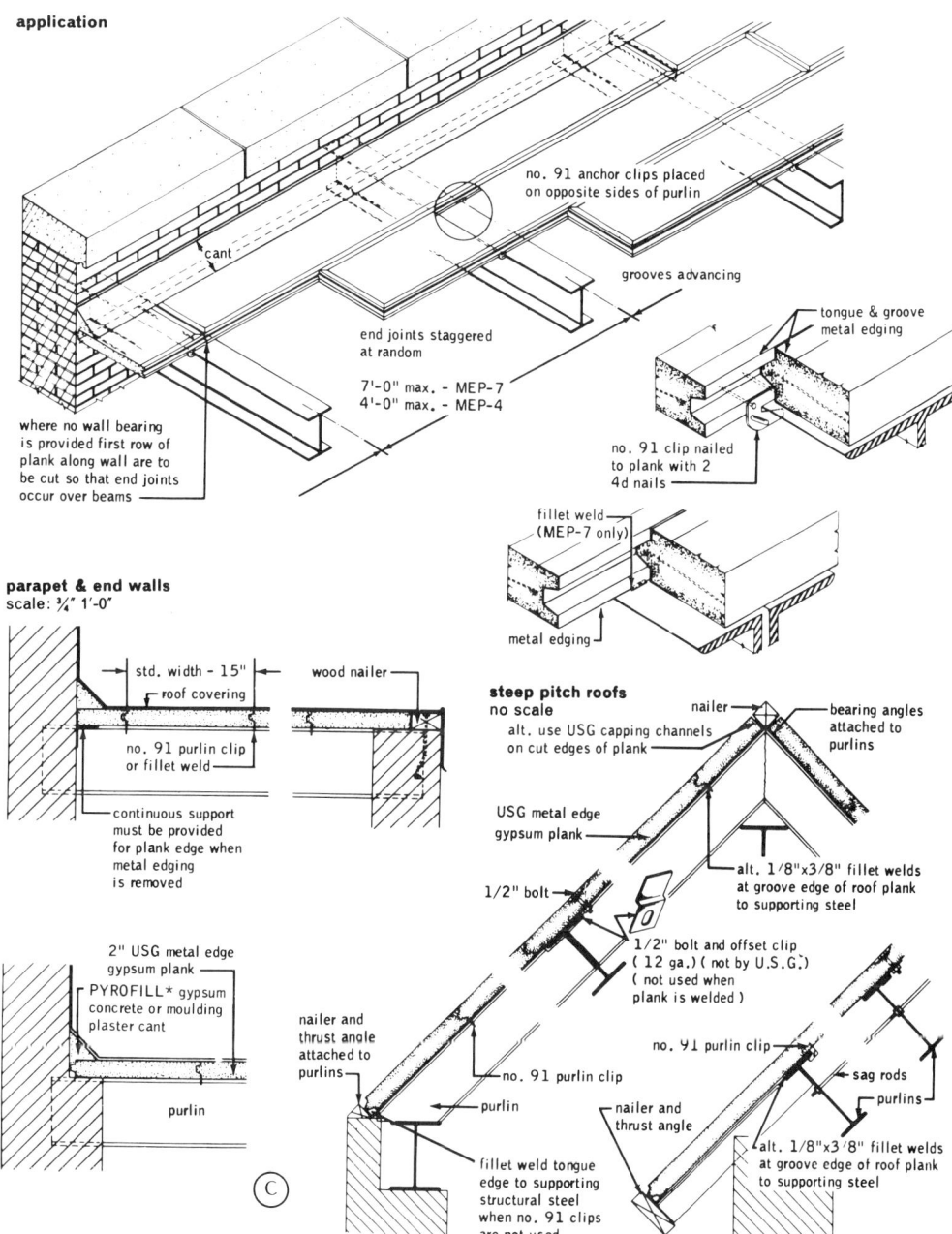

Figure 4.10 Architectural details of precast slabs.

45

Roof Decks and Substrates

2. Any roofing system on the market can be applied or replaced with little or no damage to the deck or inconvenience to the occupants. The use of mechanical scrapers is excluded because of the noise that is transmitted through the concrete frame.
3. Solid asphalt mopping to a dry deck primed with asphalt eliminates wind loss.
4. Lightweight insulating fills for drainage are possible under favorable drying conditions, or the structural slab can be sloped to drain.
5. Adequate support is provided for the 15- to 20-lb per square foot extra dead load of a ballasted protected membrane system.
6. Electric conduit can be incorporated in the roof slab or attached to the underside. False ceilings and piping can be suspended from the slab.
7. Resistance to fire or fire spread is high.

Important Considerations

Poured concrete decks present few problems if they have been correctly designed, placed, and cured. Since the coefficient of thermal expansion of normal dense concrete is 6×10^6 per degree Fahrenheit, a temperature change of 80°F will expand concrete 0.05 percent or 0.06 in. in ten feet or 0.6 in. in 100 feet. For most concrete slabs roofed, but not insulated, an expansion joint would be advisable in this distance or less, with a suitable break in the roof membrane at the same location.

Effective thermal insulation located on the cold side of the slab would reduce this degree of movement permanently, provided the insulation stayed dry and maintained its full thermal resistance value to excessive solar heat gain and a roof temperature that could reach 160°F and with an expansion of 0.96 in.

Generally, there is some shrinkage of concrete during curing and some dimension changes due to moisture content alteration; therefore roofing should be delayed until the concrete is fully cured.

In high humidity areas, penetration of moisture to reinforcing bars can cause corrosion of the steel by an electrochemical process. Iron in the presence of a liquid has a tendency to revert to its stable state, iron oxide. In the presence of oxygen and water, the reaction produces hydroxyl ions which, with the ferrous ions, form a precipitate consisting of ferrous hydroxide. If exposed to oxygen, this is converted to ferric hydroxide, the familiar reddish brown rust. Cracks in the concrete may form because the iron oxide corrosion products occupy a larger volume than the original metal. In addition, debonding of the steel may lead to breakdown of the matrix.

This emphasizes the need to protect concrete from outside precipitation and from excessive moisture vapor inside a building. Proper placement of the steel reinforcing is essential, so that it is adequately covered; 2 in. on top and 1 in. on the bottom. Concrete deflects through plastic flow or creep and this must be added to the live load deflections or elastic strain.

Concrete decks have the desired mass to mitigate rapid temperature changes. The application and removal of roof coverings and insulation is simple compared to other types of roof deck, partly because heavy equipment can be used without danger to the deck or allowing debris to filter through the deck into the building.

Lightweight concrete, perlite, pumice, or vermiculite fills are not recommended in order to achieve drainage, unless all shrinkage has taken place and the water content has dried out completely before the roof membrane is laid. Perfect weather conditions are necessary. Most of these fills pick up moisture from inside the building over time.

Precast Concrete

Dense reinforced concrete slabs installed on simple supports (i.e., single spans up to approximately 20 ft) are sometimes difficult to roof over. The following deficiencies are apt to occur:

1. Side joints do not line up.
2. Slabs are often not keyed together on the same plane.
3. Thermal movement is reflected in the roof covering at the end joints of the slabs.
4. Plastic flow causes deflection between the supports.
5. Insufficient or improper curing adds to item 4.
6. Concrete is usually so dense that it makes hand drilling or gun-fired fastening of roofing components very difficult.
7. Felts, base sheets, or vapor barriers cannot be secured with continuous moppings of hot bitumen because of possible drippage through the joints.
8. Firm attachment of PVC vapor barriers and insulation with cold adhesives cannot be guaranteed because of uneven surface planes between the slabs (see Figure 4-11).
9. Some lightweight channel roof slabs designed to span 27 ft are too springy (see Figure 4-12).

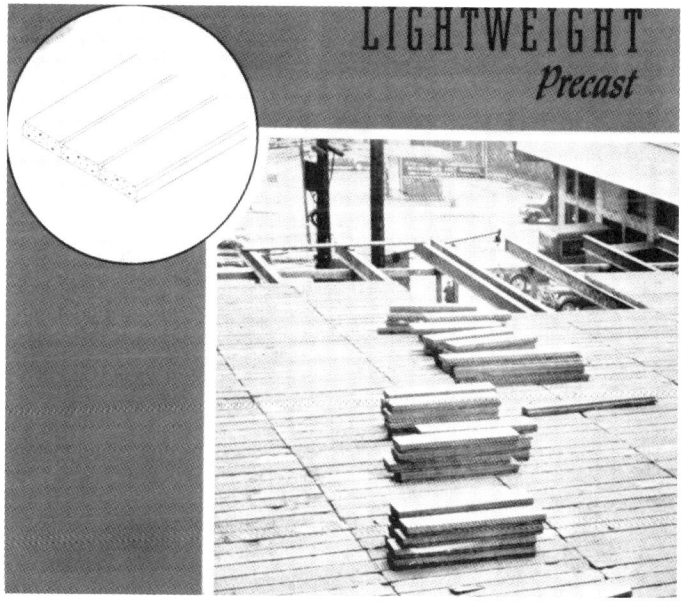

Figure 4.11 Concrete roof slabs. The neat drawing on the left does not agree with reality.

ROOF SLAB

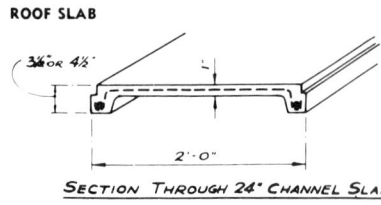

SECTION THROUGH 24" CHANNEL SLAB

Width Of Slab	Thickness Of Slab	Max. span for Flat Roofs	Weight per sq. ft. in lbs.
2'-0"	3½"	8'-0"	14
2'-0"	4⅝"	12'-0"	16
1'-0"	4½"	16'-8"	23

FLOOR SLAB

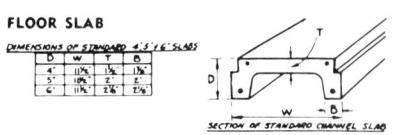

Thickness Of Slab	Weight per Ft.	Span in Feet						
		8	10	12	14	16	18	20
4"	22	250	155	110				
5"	30	420	310	200	160			
6"	36				260	200	160	125

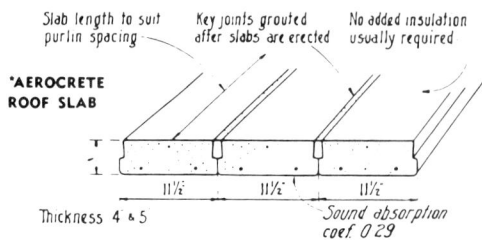

*AEROCRETE ROOF SLAB

STANDARD AEROCRETE ROOF SLAB DESIGN DATA
SNOW LOAD 40 Lb. Sq. Ft.

Thickness of Slab	Maximum Span for Flat Roofs	Weight per square foot pounds	Thermal Conductivity per inch thickness	Heat Transmission through complete roof with tar and gravel and allowing for 15 M.P.H. wind
4"	8' 0"	23	2.0	0.31
5"	11' 0"	29	2.0	0.27

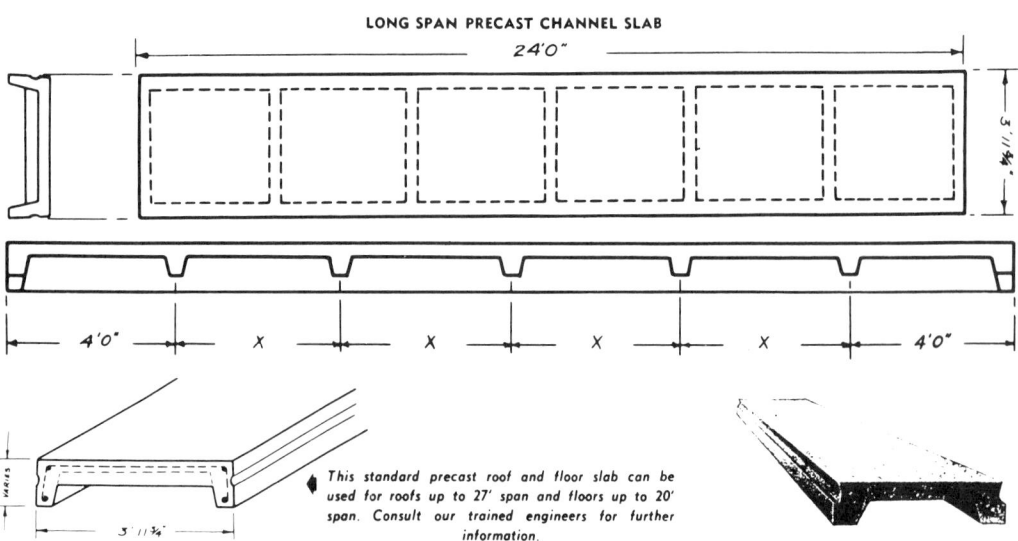

Figure 4.12 Experience with channel slabs shown is similar to slabs in Figure 4.11.

10. Most precast slabs have little or no thermal resistance and minimum ability to absorb moisture without damaging the steel reinforcing.

11. Lightweight slabs made with special aggregates have some advantages over dense concrete slabs, but they must be selected with care so that roofing application and performance problems are reduced to a minimum. Manufacturers should be consulted.

12. A 2-in. concrete cover reinforced with welded wire mesh can be used to cover the slabs and provide a suitable base for roofing. This must be dry and keyed or bonded to the slabs. Allowance should be made for movement in the areas of greatest stress to prevent splitting of the roof covering.

Precast Concrete – Long Span

Prestressed concrete roof decks may be double tees, single tees, F-shaped, or cored slabs (see Figure 4.13). Spans up to 70 ft are possible with double-tee units 8 and 10 ft wide. They are held together with a welded joint. A flat bar is laid across weld angles that are cast in the flanges (see Figure 4-14).

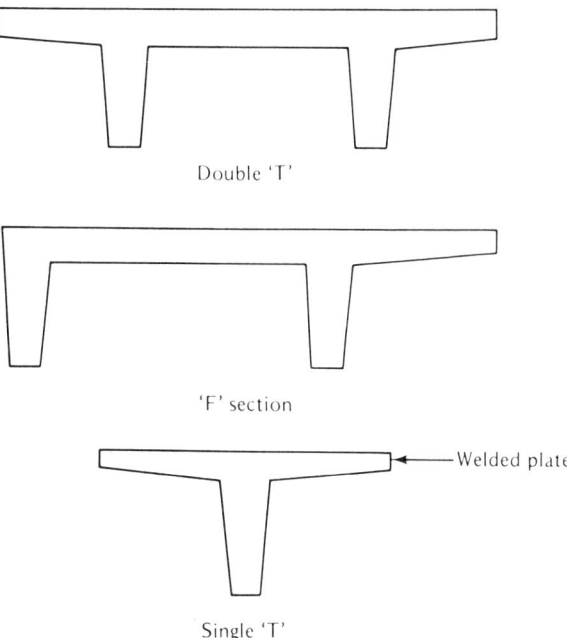

Figure 4.13 Long-span precast concrete slabs.

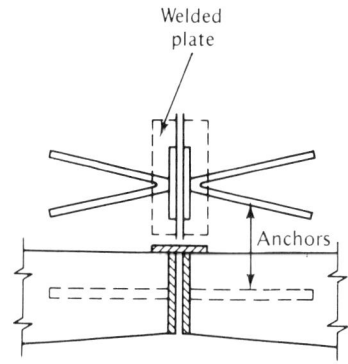

Figure 4.14 Joining long-span precast concrete slabs.

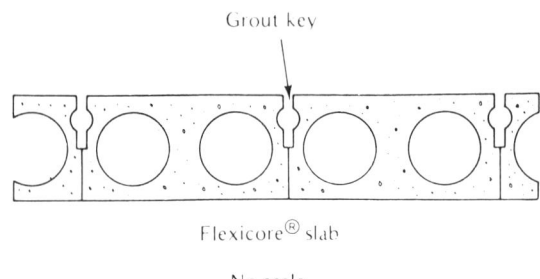

Figure 4.15 Grouting long-span precast concrete slabs.

Cored slabs are grouted together with a grout key and tied across a beam with reinforcing bars (see Figure 4-15).

Sloped decks are possible by cambering the slabs, varying the height of columns and beams, or covering with a lightweight concrete topping or dense concrete.

Roofing problems arise if there is thermal movement of large concrete sections. The coefficient of thermal expansion per degree Fahrenheit for normal dense concrete is 6×10^{-6}. The deformation due to a temperature change of 80°F is 0.05%, or 0.06 in. in 10 ft. In a 70-ft-long section 10 ft wide, the dimensional change through 80° would amount to 0.42 in. in the length and 0.06 in. in the width. Where two sections come together these amounts can be doubled. The coefficient for steel is 7×10^{-6}, which makes it an ideal material for reinforcing concrete. The modulus of elasticity E for concrete is 2.5×10^{-6}; for steel it is 30×10^{-6}.

These dimensional changes in dense concrete make it advisable to keep the concrete in a stable temperature environment and if possible separated from the continuous roofing membrane. Control joints in any concrete structure should be carefully located and waterproofed with adequate caulking materials or fabricated expansion joints.

Important Considerations

1. Precast concrete, both dense and lightweight, may have some advantages over other decks in certain circumstances but these advantages are in construction and not in roof membrane design or performance. All of the precast concrete decks with which the author has had experience have presented difficulties that were hard to overcome.

2. Long-span prestressed concrete roof decks are also difficult to roof over because of the thermal movement, jointing, and springiness. Prestressed or post-tensioned T-slabs are much more flexible than ordinary reinforced concrete poured in place. Flexible roof decks may prove to be suitable for equally flexible synthetic rubber sheet materials as long as joints in the deck are properly bridged and provided the roof material possesses adequate rheological properties, i.e., the capacity to bridge cracks without cracking or necking. This could be a problem after aging has reduced the original elasticity.

Lightweight Concretes

Structural lightweight concrete is defined as concrete that has a 28-day compressive strength of 2,500 psi, and an air-dry unit weight of less than 115 lb per cubic foot. This type of concrete should not be confused with very lightweight concretes used primarily for insulation purposes. Such concretes have a unit weight range of 15 to 90 lb per cubic foot. Their compressive strengths seldom exceed 1,000 psi.

These concretes may be grouped as follows:

Group 1: Those made with aggregates of expanded materials such as perlite or vermiculite.

Group 2: Those made with aggregates manufactured by expanding, calcining, or sintering materials such as blast furnace slag, clay, fly ash, shale, or slate; or by processing natural materials such as pumice, scoria,[4] or tuff.[5]

Group 3: Those made by incorporating in a cement paste or cement sand mortar a uniform cellular structure of air voids by using preformed or formed-in-place foam.

Drying shrinkage: Unit weight in the plastic state is 90 to 120 lb per cubic foot made and cured at normal temperatures. The shrinkage is generally slightly greater than normal-weight (149 lb) concrete. The difference in shrinkage is usually less than 30%; in some cases there is little or no difference.

The shrinkage of insulating concrete is not usually critical when it is used for insulation or fill, except that excessive shrinkage can cause curling. For structural use, shrinkage should be considered. Moist-cured cellular concretes made without aggregates show high shrinkage. Moist-cured cellular concretes made with sand may shrink 0.10% to 0.60%, depending on the amount of sand used. Autoclaved cellular concretes shrink very little. Insulating concretes made with perlite or pumice aggregates may shrink 0.10% to 0.30% in six months at 50% relative humidity. Vermiculite concretes may shrink 0.20% to 0.45% in the same period. Shrinkage of insulating concretes made with expanded slag or expanded shale ranges from about 0.06% to 0.11% at six months.

Expansion joints: Some producers of aggregates recommend a 1-in. expansion joint at the juncture of the concrete and all roof projections. Transverse joints are used at a maximum of 100 ft in any direction, for a thermal expansion of 1 in. per 100 lineal feet. A fiberglass material that will compress to one half its thickness under a stress of 25 psi is generally used.

Certain types of insulating concretes may not require expansion joints because the minimum initial shrinkages are greater than the maximum thermal and moisture expansion that may be expected. It will be seen from the above brief descriptions that

---

[4]Scoria: a cinderlike basic cellular lava.
[5]Tuff: fragmented rock consisting of the smaller kinds of stratified volcanic detritus.

Roof Decks and Substrates

lightweight concretes need to be thoroughly understood and handled with care as a roof base.[6]

Table 4.1 indicates the preferred method of attachment for low-density substrates.

Wood Fiber and Cement Slabs

Structural cement-fiber roof decks are composed of wood excelsior (pine or aspen), portland cement, and calcium silicate compressed into flat slabs. Edges can be square, T&G, or rabbetted. Dimensions are $22^{1}/_{2}$ to $46^{1}/_{2}$ in. wide and 48 to 192 in. long. Thicknesses vary from $1^{1}/_{2}$ to $3^{1}/_{2}$ in. The density is approximately 22 lb per cubic foot or 4 lb per square foot 2 in. thick. Thermal resistance R is approximately 1.75 per inch for unsurfaced sheets. It is also made 2, $2^{1}/_{2}$, and 3 in. thick with up to $1^{1}/_{2}$ in. of urethane foam insulation bonded to the upper surface. The highest thermal resistance R is reported to be 16.37 including the roof covering and air films. The U value is 0.06.

Spans are increased up to 6 ft for 3-in.-thick material when 16-gauge steel channels are inserted in the T&G edges. Panels are secured with clips, screws, and special nails.

Note: When lock-type fasteners are used to secure felt, base sheet, or insulation to wood fiber cement roof decks, the roof can be removed without destroying the deck. However, when urethane insulation is bonded to the base material and the roof membrane is mopped to the insulation, the membrane cannot be removed without destroying the entire roof deck. It might also be difficult to resurface an old roof because of the violent action that must be taken to remove old gravel and make other repairs.

Manufacturer's technical literature should be consulted for other important design details.

Asbestos Cement Cavity Decks

Asbestos cement cavity decks are built up from units or sheets 3 ft $7^{1}/_{8}$ in. by 10 ft by $^{3}/_{8}$ in. thick (109.55 by 304.80 by 0.97 cm). Individual sheets are formed as in Figures 4.16 and 4.17. They are assembled as in Figure 4.18 and fastened together as in Figure 4.19. The overall depth is $4^{11}/_{16}$ in. (11.91 cm). Weight per square foot laid is 10.8 lb (4.86 kg). The thermal resistance R of deck and roof covering is 2.48. Units are clipped or bolted to steel framing.

Owing to the $^{3}/_{8}$-in.-deep depression that occurs at 15-in. centers at the side laps and the open end butt joints, it is not practical to lay a roof membrane directly on the deck; therefore, a layer of thermal insulation over a vapor barrier, if necessary,

---

[6]Complete information is included in Chapters 13 and 15, in *Design and Control of Concrete Mixtures*, Portland Cement Association, Skokie, Illinois.

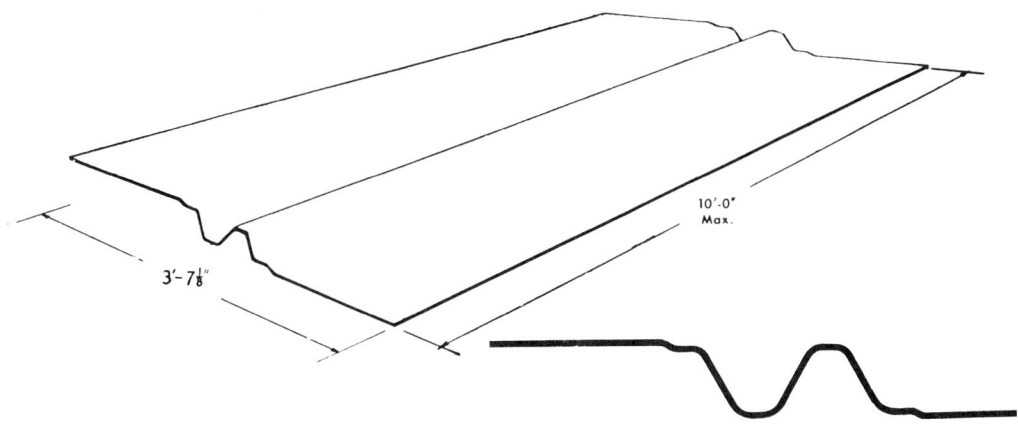

Figure 4.16 Forming individual sheets for asbestos cement cavity decks.

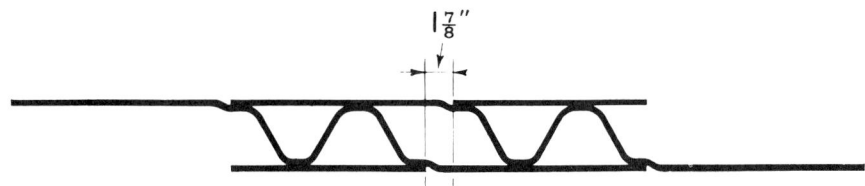

Figure 4.17 Forming individual sheets for asbestos cement cavity decks.

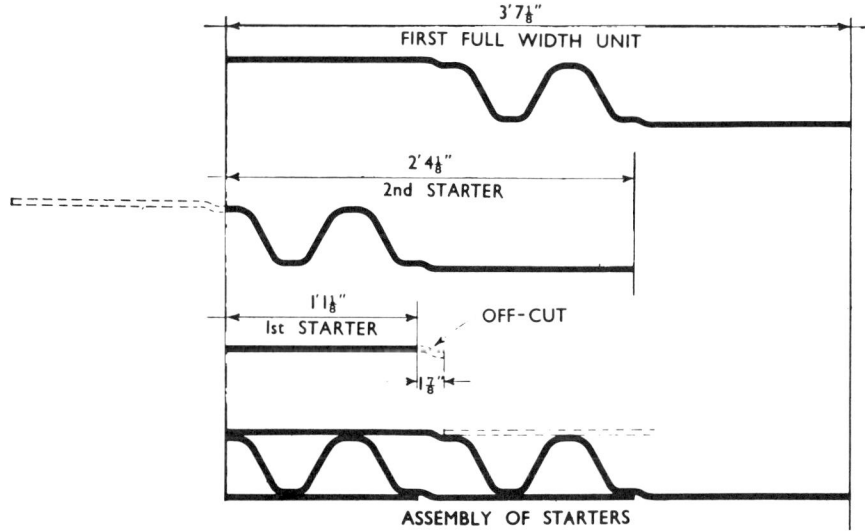

Figure 4.18 Assembling sheets for asbestos cement cavity decks.

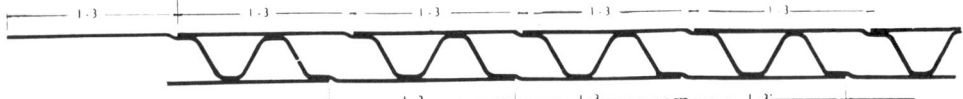

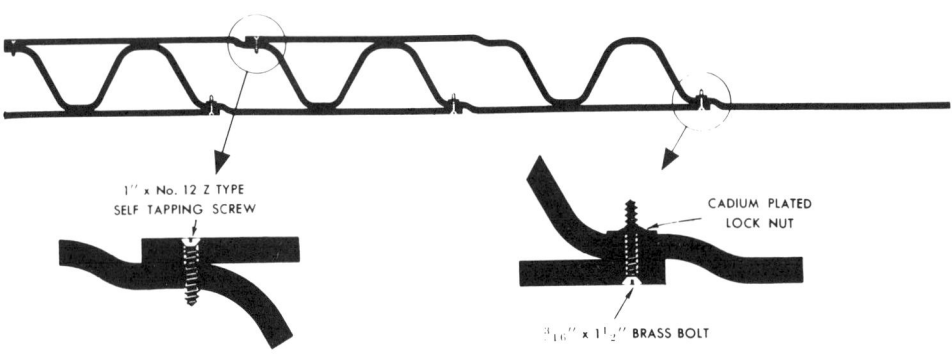

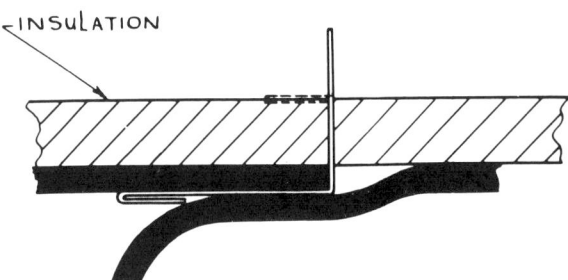

Figure 4.19 Fastening sheets for asbestos cement cavity decks.

must be secured to the deck with a suitable adhesive that will not flow or drip through the laps and joints. A low-density insulation like fiberglass may also be secured with metal cleats, as shown in Figure 4.19. These will act as thermal bridges and may be undesirable in some buildings.

Reroofing procedures might require modification if the asbestos cement units suffer embrittlement by leaching of the cement in humid atmospheres. Reference to manufacturer's technical literature and general recommendations for correct usage is essential.

5

Recommended Minimum and Maximum Roof Inclines

Authorities in several countries agree that a roof should be inclined or sloped to drain water to internal or external roof drains or overflows. In some countries a small incline is mandatory, which indicates it is not impossible structurally to provide a slope. There are few, if any, advantages to the roof membrane in deliberately designing a roof system to hold water. So-called level decks constructed for architectural and engineering simplicity do not stay level, but deflect between supports owing to live and dead loads and creep or plastic flow. All deck materials are affected to some degree.

An incline of not less than 2% is accepted as a practical minimum. This is ¼ in. in 1 ft or 2 cm in 1 m.

Problems with Level Roofs

Problems associated with level roofs include the following:

1. Water ponding forms a reservoir that continually aggravates insignificant faults and gradually makes them worse.
2. Water ponds penetrate accidental punctures or faulty flashings and fill all voids in insulated roof systems. Vapor barriers may retain the water until the entire roof is floating. This can occur almost immediately after the roof is laid. Venting will not dry out the system.

Recommended Minimum and Maximum Roof Inclines

3. Ponded water destroys all ferrous metals in flashings in a few years unless they are extremely well galvanized, painted both sides, factory enameled, or kept above the highest water level.

4. Ponded water can stain the building exterior by running over the edges where there are no gutters.

5. Ponded water can run over base flashings at walls when the design or workmanship is defective or where there is deck settlement or shrinkage.

6. Ponded water can cause additional deflection of the roof deck, which results in more water and more deflection. Water weighs approximately 5 lb per square foot for each inch of depth.

7. Mechanical equipment such as air conditioning or exhaust fans may be installed on the roof in locations that contribute to deflection not anticipated in the original design.

8. Unless drain flanges are recessed below roof level, the extra felt stripping covering the flange and the gravel stop on the drain will raise the outlet as much as $1/2$ in. This can be added to the normal deflection of the deck, unless the drain is located in midspan.

9. On some buildings, the normal shrinkage of the frame (wood or concrete) will raise the drain sump enough to fracture the roof membrane, or at least increase the depth of ponded water.

10. Fly ash and other airborne debris can block drain strainers, causing a backup of water, which can run over flashings if overflow outlets are not properly placed.

11. Ponded water holds airborne debris, which slowly floats to drain outlets, eventually blocking them. Sloped roofs flush themselves more readily. Drains are often located at columns or beams, which means the water that accumulates in deflected areas between the columns does not reach the drains. If a prestressed double-tee concrete slab without camber has a span of 60 ft and an allowable deflection of $1/240$ of the span, 3.0 in. of water can accumulate. If it is designed for $1/360$ of the span, water can reach a depth of 2.04 in.

12. Ponded water on some roofs attracts water birds at certain times of the year. They leave feathers and deposit undesirable materials that clog drains and create odors.

13. Leaks in level ponded roofs are difficult to trace and expensive to repair, especially in winter when there is ice and snow on the roof. The ice and snow can be floating on water.

14. Some damage may be done by photo oxidation, removal of water-soluble constituents in bitumens, and freeze-thaw cycles, but these are not serious hazards.

15. Smooth-surfaced asphalt roofs should not be laid on roof decks with inclines of less than $1/2$ in., and for selvage-edge cap sheet roofs, not less than 1 in. per foot. Only hot asphalt can be used for the adhesive on these minimum inclines. Low pondable areas must be avoided.

16. For the last 50 years, only roofs with multiple plies of felt and hot bitumen have been used for roofing over decks that do not drain. Cold application roofs with

two or three plies of heavy (50 lb) coated felts and cut-back asphalt cement have been used on slopes of 1 in. per foot or more. Cold-applied roofs have been surfaced with emulsified asphalt coatings, mineral granules, or a mineral surfaced roofing sheet (approximately 90 lb 100 ft^2).

After this long experience it is difficult for manufacturers of traditional roofing materials and for roofing contractors to accept a single-ply roof membrane with 2-in. laps sealed by solvent welding, heat sealing, or contact adhesive.

General Rules on Roof Inclines

As a general rule, roof gutters, valleys, canals, ditches, underground storm sewers, water pipes, and other water-carrying facilities are graded to ensure that the water flows by gravity to an outlet, unless it is pumped. A roof also serves as a water-carrying device and is not much different from the other forms mentioned. To make it dead level and place the building in jeopardy seems to be somewhat out of tune with the accepted practice of providing a suitable gradient.

On steeper inclines, the use of hot materials is dangerous, and the hot asphalt cannot be mopped or spread evenly or the felt rolled in as on low-sloped roofs. Back nailing felt becomes necessary above 1 in. per foot in most climates, which requires a nailable deck. Nailing strips on nonnailable decks are not practical.

Coal-tar-pitch roofs should not be used on inclines above $1/2$ in. per foot, and even at this slope the pitch will often flow. Slopes at roof edges should be carefully avoided on pitch and gravel roofs.

On any incline above 3 in. per foot, a monolithic or continuous roof membrane is not necessary. There are more practical forms of overlapping sheets that are watertight owing to gravity carrying the water down the roof.

Variable slope roofs such as the barrel type (convex) or the dished type (concave) are difficult to roof with hot materials.

For examples of ponded roofs, see Figure 5.1.

Figure 5.1 Examples of ponded roofs. All pose a threat to the owners and the occupants.
(A): Airborne debris blocks drains so that roof does not drain.

Figure 5.1 (cont.)
(B), (C): Same building. Drains became blocked with sawdust from neighboring mill and water ponded to a depth of 5 in. before it was accidentally discovered—25 lb ft^2.

(D), (E): These are apartment buildings and all of them are ponded most of the time. The quality of materials and workmanship on these buildings is not the best; therefore the danger of leakage is high. The lack of proper vapor barriers and effective ventilation has caused many of the decks to collapse within five years or so.

Roof Traffic

The closer a roof deck is to dead level the more activity it seems to attract. A great number and variety of people seem to have business on the roof, no matter what kind of building it is. Groups of retail stores change their type of occupancy so that roof-mounted equipment is added or subtracted. Such things as air conditioning units, air vents, television aerials, flood lights, Christmas decorations, and various sales promotion schemes all find their way to the roof and are dropped, nailed, bolted, or sandbagged to the roof membrane and to the metal flashings. Generally there is little control or no control of such activity by people who have no idea, or concern, for what they are walking over or how much damage they may be doing.

Industrial buildings have an equally hard time with the many changes in manufacturing processes that demand new equipment be mounted on the roof or moved to another location. Spillages of toxic liquids or even steam or hot water take their toll on roofing.

On some buildings, sign installers and window washers, two of the most common roof destroyers, require watching, the first with their strange sign bases and guy wire anchors, and the second with their roof hooks and window washing chemicals. However, these people are amateurs compared to school children who are able to climb around all obstructions. Barbed wire and heights mean nothing to children. Bicycle racing is a favorite sport on school roofs, together with lifting roof strainers from drain outlets and cutting off lead counter flashings from plumbing vents. The loss of a roof strainer can cause havoc when a tennis ball floats across a level roof and down the drain, to block off the water flow. If a lead vent flashing is cut off near deck level, a leak is bound to occur. All that is needed is a pen knife.

Any roof covering laid over low-density thermal insulation is in a precarious position because of all this unwanted activity. This is only one reason for the removal of insulating materials from below the roof membrane. Without insulating materials, a roof membrane is better able to withstand roof traffic and the installation of additional equipment. It is not easy to make a watertight connection to a cut-out portion of roofing when insulation is involved. The lap is over soft material.

It takes only one puncture in a level roof membrane to fill the insulation layer with water, either the material itself, or in the joints between the sheets or slabs. A vapor barrier may prevent the water from escaping, or if there is none, there will likely be moisture in the system from a heated and air conditioned interior long before the roof starts leaking.

Roofing Examples

Example 1
An addition to a large department store was roofed with a four-ply asphalt felt and gravel roof over 1-in. fiberglass insulation, a mopped felt vapor barrier on a level concrete deck. The roof held a great deal of water after a rain. The employee cafeteria opened on to this roof and some women with spike-heeled shoes walked on to the roof for a view of the city. The heels easily penetrated the roof that rested on a soft substrate, and after later rainfalls the ponded water ran into the holes and spread through the insulation layer. The whole system had to be replaced and dense fiberboard insulation used. Restrictions were placed on roof traffic and wood walkways constructed.

Example 2
A hospital with the same roof system as the department store, and the same, if not worse, ponding problem, had a two-story elevator penthouse on which a large Christmas tree was mounted. The hospital maintenance staff lowered the tree by dropping it butt first on to the roof below. They did nothing about the large hole made in the roof, so again, the ponded water ran into the insulation. If the roof had been sloped, there would not have been any water on the roof. If the insulation had not been there, the roof would not have been punctured.

Recommended Minimum and Maximum Roof Inclines

Asphalt Softening Points and Roof Inclines

Above 3 in. per foot hot-applied roofs become impractical, especially when nailing is difficult. Low softening point bitumens for BUR, i.e., below 135°–176°F (Type 1 and 2 asphalt) have a tendency to slide or flow. Higher softening point asphalt, 185°–225°F (Type 3 and 4) are also made for slopes above 2 in. Asphalt for shingles on steep slopes use an asphalt that is filled with mineral filler making it more stable when heated by the sun. Asphalt for BUR does not contain fillers.

Steep Roofs

In commercial and industrial work steep-sloped roofs such as saw-tooth skylights can be covered with steel, copper, aluminum, or terne-plated steel. These roofs can be applied with no horizontal jointing and no exposed fasteners. A safe minimum slope is around 3 in. per foot.

Where good longevity is required there is no good reason to drop down to the cold-applied asphalt-coated roll roofing that the asphalt roofing industry started with in the early 1900s. The maintenance cost is high.

Danger from Chemicals

If for no other reason, roofs should drain easily and completely to remove all traces of airborne chemicals, including nitric and sulfuric acid formed by the action of sunlight on sulfur dioxide (SO_2) and nitrogen oxide (NOx). Oil- and coal-burning thermal electric generating plants, industrial plant emissions, and the automobile are the principal villains, but such natural processes as volcanic eruption, forest fires, and the bacterial decomposition of organic matter can produce the acid sulfur and nitrogen compounds that form acid rain.

While asphalt and pitch bitumens are relatively inert chemically, the felts they are combined with can be adversely affected. Low-density foam insulation materials often built into the roof system and sometimes placed above the membrane in the Protected Membrane System are vulnerable. Concentrated acid solutions left on dead level roofs could cause a great deal of damage. Metal flashings and even stone can be affected.

> The pH scale is scientific shorthand for measuring acidity or alkalinity. The scale ranges from 0 to 14. A neutral solution is pH 7. Because pH values are logarithmic, pH 1 is ten times as acidic as pH 2, 100 times as acidic as pH 3, and so on. Thus the worst acid rainfall measured in the United States—pH 2.3 at Kane, Pa. in 1978—had about 1,000 times the acidity of pure rain, pH 5.6.[1]

Airborne chemicals in various cities can be determined by checking with local health or environment authorities. Even in rural areas, acid rain is present if reports of dead lakes are to be believed; there is plenty of evidence that this is so. Serious

---

[1] Russ Hoyle, "The Silent Scourge," *Time*, November 8, 1982, p. 99.

damage to lakes has been reported in Massachusetts, Maine, New York, Minnesota, Wisconsin, Florida, California, Ohio, Pennsylvania, West Virginia, and Ontario, Canada.

It is too early to tell what damage will be done, and how quickly, to the newer single-ply roof membranes made from PVC, elastomeric materials, rubber compounds, and modified bitumens.

The increasing amount of airborne wastes that can today deposit 92 tons per square mile per month, with peak amounts as high as 200 tons, surely calls for a sensible approach to roof drainage and an extremely careful selection of roofing materials and systems. Dead level roof decks and haphazard drainage can only result in more serious failures. Protected membranes that cannot be readily flushed clean under the insulation and the ballast may suffer from air contaminants. Chemicals in rain water filtering through the insulation layer may be the downfall of foam insulation materials.

Moss

Mildly acidic water ponding occurring casually or intermittently on level roofs can produce a healthy growth of moss in damp climates. The moss can grow into the roofing felt if the top coating is cracked or alligatored and it can dislodge the flood coat and gravel. Grass and small trees can grow in the moss to produce even more damage due to the longer spreading root growth. Since it is the gravel that catches the moss spores it is wise to choose a smooth surfaced roof on an inclined deck that can be drained and flushed.

Using alkaline sprays or lime to change the pH from acid to alkaline is not effective over the long term because they are soon washed away. Small pieces of lead or zinc may possibly reduce moss growth on some flat roofs. It is the oxides of these metals that kill moss and discourage new growth.

What To Do About Dead Level Roofs

It is easy to explain why level roofs should never be built, but they are a fact of life that will not go away. Thousands of squares of roofs are flat and will stay flat. The disagreeable and damaging features will remain, unless of course they can be reduced or removed entirely. The following possibilities should be examined.

1. Install more drains to reduce ponded water. Note where ponding occurs and check the strength of the deck.
2. Clean the roof thoroughly and add a second pour coat of hot bitumen and new gravel.
3. Check all metal flashings, drains, parapet walls, and penthouses to make sure there are no leaks from these sources.
4. Install automatic syphon draining devices to keep ponded water less than 1 in. deep.
5. Install walkways for servicing equipment on the roof. Concrete pads are preferable to wood walks because they do not rot and do not float away.

Recommended Minimum and Maximum Roof Inclines

6. Monitor all roof traffic and keep records of who had access to the roof and what work, if any, was carried out. Check the work when completed.

7. If moisture detection devices or cut-outs reveal moisture in an insulated system, remove the entire roof system and replace it with something that will work better. On dead level decks, such a system may be hard to find. If the insulation can be left out, that might be a step in the right direction, but if that is done the temperature gradients in winter and summer need careful scrutiny to determine possible thermal movements in the deck, and condensation of vapor on the under side in winter.

8. Cover the existing roof with a sloped metal roof if the building shape and size and weather conditions permit. Consider the cost of this compared to the cost of replacing the old system every few years. Some savings in energy costs might be made by installing new insulation on top of the old roof.

9. The author has a personal objection to the use of varying thicknesses of tapered insulation over an old roof to create a slope. It not only causes a varying thermal resistance across the roof, but it increases the danger of water accumulation under the roof membrane. We have been trying to eliminate the insulation, not add to it.

10. Break up large uninterrupted roof areas with control joints to reduce the possibility of shrinkage cracks and damage to perimeter flashing.

11. Large roof areas in a single plane can be broken up with skylights or monitors running the full width of the building. Many old factories were built in this manner. Instead of air conditioning and other mechanical units such as air-to-air heat exchangers or heat pumps or ventilating fans being mounted on the roof, they can be installed in the raised sections. The fewer the penetrations of the roof deck and roof membrane, the better. They can have flat or slightly sloped roofs or a sawtooth configuration with the sloped side roofed with metal or glass, or be used as a solar collector for space or water heating. The vertical side can be walled in, glazed, or constructed with automatic-opening smoke vents in case of fire. Monitors can also be a simple triangle with two slopes at a 45° angle with the horizontal.

12. Consider the roof deck and structure as something more than just a system to carry a little rain and snow. It does a great deal more and needs more strength and durability built into it.

6

Product Description and Physical Properties

Asphalt

Asphalt for hot built-up roofing is identified by the softening point range, with the packages or containers labeled Types 1, 2, 3, and 4 in the United States and Types 1, 2, and 3 in Canada (see Table 6.1). Softening point is affected by the length of time the asphalt flux remains in the blowing still. The softening point is not the melting point, but merely a measure of flow under controlled conditions. With most asphalts an increased softening point brings about a reduced resistance to weathering; however, using a soft asphalt where elevated ambient temperatures are the rule might cause other more serious sliding or ply separation problems.

Asphalt is a complicated material, both chemically and physically, and is not evaluated by softening point alone, although this and the viscosity may be the most important aspects as far as the designer and roofer are concerned. Other properties that become part of the total evaluation are ductility, penetration at various temperatures, flash point, burning point, loss on heating, penetration of residue, and solubility in trichloroethylene. These properties are set forth in the specifications published by the American Society for Testing and Materials (ASTM) and Canadian Standards Association (CSA).

Owing to the different behavior characteristics of asphalts from different oil

Product Description and Physical Properties

Table 6.1

	Asphalt Type			
	1	2	3	4
United States Softening Point range (ring and ball)	F, 135–150 C, 57–65	160–175 71–79	180–200 82–93	205–225 96–107
Canada	F, 140–150 C, 60–65	165–175 74–79	190–205 88–96	— —

C, Celsius; F, Fahrenheit

fields, roofers must be careful in their sources of supply and be alert to any changes noted by the roofing crews. A haphazard method of buying asphalt and felt according to price might easily lead to problems arising out of incompatibility of asphalts from different sources. Incompatibility can occur between mopping asphalts and felt saturants or between different types of roofing felt.

Research has been under way for several years to determine the feasibility of employing viscosity grading of roofing bitumens (both asphalt and pitch) as the primary criterion for the construction of the composite membrane. This means that viscosity will relate to temperature in the kettle and at the mop in order to obtain the optimum mopping thickness between the felt plies. It is assumed that the appropriate information on the equiviscous temperature range will be carried on the containers and shipping notices. It is also assumed that the roofer knows the correct bitumen temperature at all times. For additional information relative to application procedures, refer to chapters 14 and 15.

Properties of Straight-run or Unfilled Asphalt

1. Dark brown to black hydrocarbon mixture, which can be gaseous, liquid, semisolid, or solid.
2. Ninety-nine percent soluble in carbon tetrachloride or trichloroethylene.
3. Not soluble in water (0.001 to 0.01% over long period). Under the influence of water, excessive temperature, and unprotected from solar radiation, all bitumens can be slowly broken down to carbon dioxide and water.
4. Chemically inactive or inert.
5. Flash point, 450°F (232.22°C). Variable.
6. Excellent adhesive except when heated close to or above normal softening point. Naturally occurring asphalts were used as adhesives and waterproofing as early as 3800 B.C.
7. Nontoxic when heated.
8. Identified by four softening points in the United States and three in Canada (see Table 6.1).
9. Some variations in behavior depending on petroleum source and blowing procedure.
10. Viscosity usually in the general range of roofer acceptability when heated between 375° and 450°F (190.56° and 232.22°C).

11. Convenient package sizes plus tank truck for large quantities and easier handling.
12. Wide range of uses in building construction.
13. Specific gravity, 1.01 to 1.03.
14. Higher softening point asphalts require longer heating periods to reach convenient viscosity; therefore, the quality of the material may be adversely affected in the kettle or tanker.
15. Extended periods of overheating in closed containers (tankers) may lower the softening point.

Coal-Tar Pitch

In North America coal-tar pitch has been used in hot built-up roofing for more than 100 years but is now practically nonexistent in Canada. The supply appears to be dependent on the activity of steel producers who use coke or oxygen in their smelting process. The use of oxygen instead of coke or a reduction in steel output would reduce the coal-tar pitch supply. Roofing pitch is sold in steel barrels or by tanker truck. The low-melt-point pitch and its cold flow properties make it difficult to package in paperboard containers. The softening point (cube in water) is generally in the 140° to 155°F range (60° to 68.3°C). See chapter 15 for handling and storage of materials.

Coal-Tar Pitch Properties

1. Dark brown to black hydrocarbon but not related to asphalt either chemically or physically.
2. Solubility in carbon disulfide, 65% to 85%. This indicates up to 35% carbon content.
3. Flash point, 248°F (min.).
4. Specific gravity, 1.22 to 1.34.
5. Softening point, 140° to 155°F (60° to 68.3°C), CSA A123.13 (1953). 129° to 144°F (54° to 62°C), ASTM D 450-71.
6. Not soluble in water (same as asphalt).
7. Viscosity drops rapidly on rise in temperature.
8. Exhibits cold flow properties at normal temperatures if not kept on level surfaces.
9. Develops highly toxic fumes at elevated temperatures.
10. Good wetting properties under ideal conditions, but thin films cool rapidly in cold weather before good adhesion is obtained. The felt rolls must follow the mop very closely.
11. Heating time to mopping temperature and desirable viscosity is low compared to asphalt, but volatile constituents are quickly driven off when the kettle is improperly operated.
12. Danger of fire in the kettle is high due to the low flash point of pitch. Approximately 248°F (120°C).

Product Description and Physical Properties

13. Compatibility of pitch with asphalt is questionable; therefore, it is inadvisable to combine them. See ASTM Test Procedure D 1370-58.
14. The supply may be affected by the amount of coke produced for steel mills and if coke-fired thermal electric generating stations replace oil-fired stations.

Organic Felt: Asphalt

The most common roofing felt used in built-up roofing is No. 15 in 432 ft^2, four-square rolls 36 in. wide and 144 ft long (91.44 cm by 43.89 m). It is also made in three-square rolls. The advertised weight is 60 lb (27 kg), but the actual weight can be as low as 48.80 lb (21.96 kg). This should be taken into consideration when weighing cutout samples. The fabric is usually perforated with needle-sized holes at 0.5 to 1.0 in. on centers to assist the escape of air and steam when the felt is laid. Unperforated rolls are also available for other purposes. To help the roofer in maintaining the correct lap in shingle-type application, the fabric is lined for one-, two-, three-, and four-ply construction. A 432 ft^2 roll will therefore cover 400 ft^2 one ply, 200 ft^2 two ply, 133 ft^2 three ply, and 100 ft^2 four ply. The lines are always on the inside or concave side of the fabric (see Figure 6.1).

Rolls may be wound loose or tight depending on the manufacturer and the on preferences of the customers. Loose rolls generally do not stick and therefore roll out easily and quickly on the roof; however, they must be handled more carefully so that the ends are not crushed or torn. Tightly wound rolls may stick if the asphalt saturant is not dried in and the fabric cooled before winding. Sticking is bothersome to the roofer because it slows the operation and may cause rippling of the felt, which can result in a poor roof. Such rolls should be discarded. It is essential that a constant lap be maintained and that the felt roll out straight. Occasionally, for various reasons, organic felt will not follow a straight line. When this happens it must be cut and realigned or the roll discarded.

Number 30 weight saturated felt is also made but has limited use in built-up roofing.

Although organic felts are able to absorb a minimum of 140% by weight of saturant, they are still susceptible to moisture, which causes swelling and shrinking principally in the cross-machine direction. In the dry state they are generally stronger under a tensile stress than other types, but suffer considerable reduction when wet. It is not fair to say that they rot or decay. One of the attractive features of organic felts is that they use a good deal of what might be called recycled waste materials, but need machinery and energy to convert them into useful roofing felt.

Organic Felt: Tar Saturated

About the only form of tar-saturated felt for built-up roofing is the No. 15 weight, lined, but not perforated, in 432 ft^2 rolls 36 in. wide. There are standards for heavier weights and they may be available in some areas for other purposes. Roofing felts for tar saturation can be identical to those for asphalt saturation, but a few obscure but important behavioral differences can develop.

1. The felt saturates easier and faster with coal tar, but the percentage of saturant to felt weight is the same as asphalt.

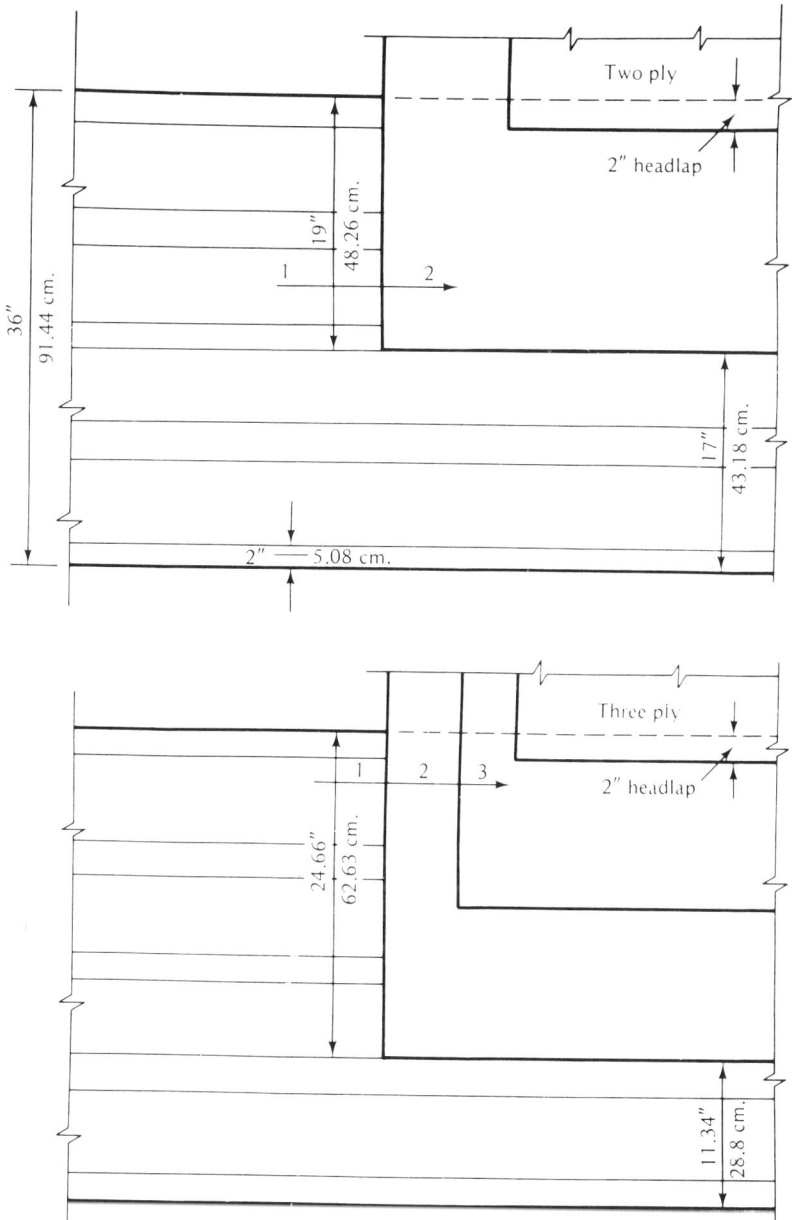

Figure 6.1 Lining and lapping 36-in.-wide roofing felt for two- and three-ply roofs.

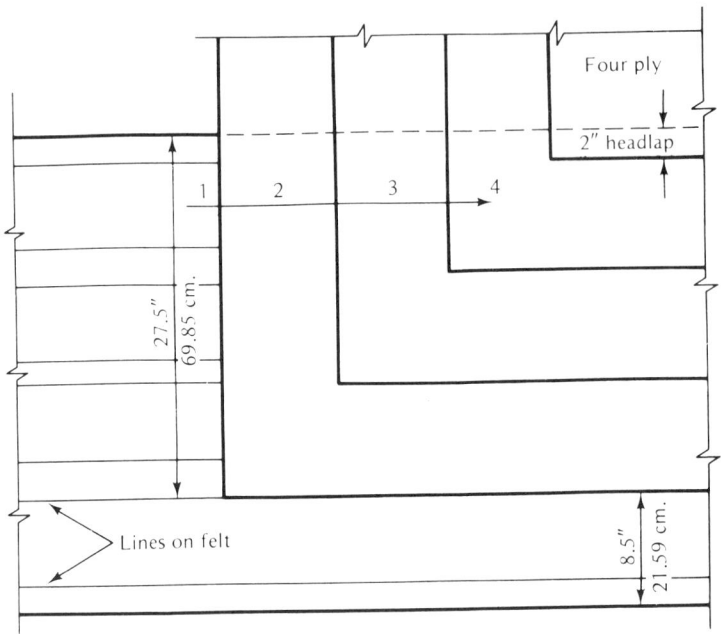

Figure 6.1 (cont'd.) Lining and lapping 36-in.-wide roofing felt for four-ply roofs.

2. Tar-saturated felt absorbs more water faster than asphalt felt.
3. Hot coal-tar pitch penetrates tar-saturated felt much more than hot asphalt into asphalt-saturated felt; therefore, constant mopping thicknesses are more difficult to obtain.
4. Tar-saturated felts are not suitable for any construction use other than in roofs for level or near-level roof decks or for some types of underground waterproofing or damp proofing.
5. Tarred felts, even with a high rag content, lose their pliability in built-up roofs after 10 to 20 years of normal exposure. This is due to loss by evaporation of the lighter fractions in the coal-tar saturant, which leaves a high percentage of carbon.

Asbestos Roofing Felts: Asphalt and Coal-Tar Saturated

Number 15 asbestos roofing felts are made in 432 ft² rolls, 36 in. wide, lined and perforated. Asbestos fiber content is 75% to 85% of the total fiber used. The balance is organic or glass plus starch binder to improve saturation. Asbestos fibers do not absorb saturant, which results in the weight of saturant being approximately 40% (minimum) of the total weight of the product, or 60 lb per roll (asphalt or tar). This means that the weight of unsaturated fabric in asbestos felt is approximately 43% more than in an organic felt.[1]

---

[1] Manville Corp. are reported to be discontinuing the sale of asbestos roofing felts, including asbestos-containing mastics, effective Dec. 31, 1983. Other manufacturers of asbestos roofing products may follow suit. This action may affect the suggested roofing specifications in chapter 10.

Being relatively free of organic materials and having lower saturation weights, asbestos felt has certain advantages.

Glass-Fiber Felt

1. The convolutions of the rolls seldom stick together.
2. Storage under unfavorable conditions is not as great a problem as with organic felts.
3. Perforations are cleaner and stay open to vent out air and steam in hot applications. Asbestos felts were the first to use the perforation principle because there was less bleeding through the felt itself.
4. Water absorption is low but not entirely eliminated.
5. The felt generally rolls out straight and is free of rippling and fishmouthing at the edges.
6. Asbestos felt roofs should qualify for better fire-resistance ratings; however, the entire system must be evaluated, not just the roof membrane alone.

Glass-Fiber Felt

The following are basic data for asphalt-saturated glass-fiber felt.

ASTM Designation D 2178-76. Asphalt-impregnated glass mat for roofing and waterproofing.

3. Classification
3.1 Asphalt-impregnated glass mats covered by this specification are four types.

			Mass	
			(g/m^2)	lb/100 ft^2
3.1.1	Type I	Utility ply sheet	49	1.0
3.1.2	Type III	Standard ply sheet	73	1.5
3.1.3	Type IV	Heavy-duty ply sheet	83	1.7
3.1.4	Type V	Combination sheet	49	1.0

CSA Standard A 123.17 (1963)

Type I	Ply sheet	7.5 lb/100 ft^2
Type II	Base sheet	14.0 lb/100 ft^2
Type III	Combination base sheet	14.0 lb/100 ft^2

Weight of desaturated mat: 0.85 lb/100 ft^2

Other types made with no CSA standard are 30-33-40 coated base sheets, and granule surfaced and perforated sheets (5/8-in. holes on 3-in. centers). This is intended as a venting base sheet that is laid granule side down on a dry deck and mopped over to button the sheet to the deck or substrate. Venting to the roof edges is supposed to take place.

It should be noted that glass felts have a glass mat weight not exceeding 1.7 lb 100 ft^2. It can be as low as 0.85 lb. The tensile strengths are more nearly equal in machine and cross-machine directions than wood fiber or asbestos felts. The actual

Product Description and Physical Properties

breaking strains are hard to pinpoint because they vary with moisture and temperature and the characteristics of the bitumen with which they are combined.

Tests show glass felts are not affected dimensionally by moisture as much as wood fiber or as asbestos, but temperature changes have a greater effect.

Owing to the changes taking place in the production of glass-fiber felts, no value judgment will be made. It is understood that production and use of some types are on the increase in the United States. Being inorganic, they have some advantages.

Smooth-surfaced Roll Roofing: Organic and Inorganic

Perhaps the first and still the most common asphalt roofing material, smooth asphalt-surfaced roll roofing, is made for several purposes by changing the basic felt weight and the weight and distribution of filled asphalt coating, which can be on one side or both sides of the sheet. Rolls can be 50 to 65 lb per 108 ft^2 (23 to 29 kg per 10 m^2) and 60 to 87 lb per 216 to 228 ft^2 rolls. All are 36 in. (0.91 m) wide.

Uses include the following:

1. Single-ply roofing with nailed and cemented 2- and 4-in. laps.
2. Base sheet for built-up roofing, nailed or mopped.
3. Vapor barrier under thermal insulation. Also termed vapor seal or vapor retardant.
4. Flashing reinforcement.
5. Two-ply hot applied roofing. Failures are reported over insulation.

> **Note:** The vapor permeance of asphalt-coated roofing is generally very low, but it can be affected by how the sheet is made, calendered, saturated, and coated, and the type and amount of filler in the coating.

Mineral-surfaced Roll Roofing

Mineral-surfaced roofing (MS) is essentially the same as smooth-surfaced roofing (SS), except that one side has a heavier coating of asphalt plus mineral granules. A 108 ft^2 roll 36 in. wide generally weighs 90 lb on an organic felt base, and 75 to 80 lb on a glass base. A roll covers one square (100 ft^2) with a 2-in. side lap and a 6-in. end lap.

Some manufacturers supply the standard MS roofing in 9-in. and 18-in. widths principally for asphalt shingle roofs, but these widths are also used in some flashing details on built-up roofs.

Mineral-surfaced, Selvage-edge Roofing

This is essentially the same as 36-in. wide MS roofing, except that only 17 in. of the sheet is covered with granules. The balance is left as a selvage edge for lapping the roofing two ply. The selvage edge and the back are generally not coated when used with hot asphalt. A typical weight per roll is 55 lb (110 lb per square). If the material is to be laid with cold cement, a lightweight asphalt coating is applied and the weight is increased to approximately 60 lb per roll (120 lb per square).

7

Thermal Insulation

When buildings are heated and cooled, some means must be incorporated in the roof design to prevent excessive heat loss in winter and heat gain in summer. Some manufacturing processes demand close control of the interior environment and of storage areas, a constant condition in both winter and summer for the protection of the finished product.

When a roof is steeply sloped, it is possible to place bulk-type insulation, either in batts, blankets, or in loose fill form, in a flat ceiling and ventilate the space above it. The same type of insulation may be placed between the roof rafters and a roof covering, and overall design arranged so that moisture will not be trapped.

Flat roofs on frame construction may also be insulated with bulk insulation below the deck, and the space above the insulation ventilated to the outside. With careful design and execution, the system is usually successful and is relatively inexpensive, principally because it seldom has to be dismantled and reconstructed owing to damage from moisture in the wrong place. In a flat-roof system, however, on most residential and commercial buildings the thermal insulation is a rigid board type located above the structural deck. It can be wood fiber, cane fiber, cork, fibrous glass, foamed cellular glass, foamed polystyrene, extruded polystyrene, foamed polyurethane, or expanded volcanic ore combined with wood fibers and asphalt binder (perlite). Some insulation boards are combinations of perlite and urethane that

Thermal Insulation

assist in the hot mopping of felts. Others, like polystyrene, have felt applied at the factory for the same purpose. Fiberglass is also combined with perlite or urethane. The insulation value of any material depends on the independent air cells rather than on the material in the cell walls. Generally, organic materials are slightly better than mineral, but they may allow a greater diffusion of moisture vapor, which, when condensed, will reduce the thermal resistance.

The published thermal insulation values k, which is the heat transfer through 1 ft^2 of material 1 in. thick in 1 hour per degree Fahrenheit difference in temperature measured in British thermal units (Btu's), or the reciprocal R, which is the resistance to heat transfer of 1 in. of material, are based on bone-dry material, which is rarely obtained in service. When making comparisons, it must be clear that the values obtained are the result of the same standard test method (e.g., ASTM C 177). The thermal resistance value R is important in order to compute the total thickness required if a life-cycle costing system is used. However, other mechanical properties of insulation materials are equally important and affect their performance. These properties should be studied very carefully if they are in the manufacturer's literature. If not, one should start asking questions. A few of these physical properties are listed below.

1. Cost per unit of thermal resistance. Divide the cost per 1,000 ft^2, 1 in. thick by the resistance R.
2. Thermal coefficient of expansion and general dimensional stability.
3. Horizontal shear strength. Resistance to wind suction.
4. Compression resistance and recovery under point and rolling loads and mechanical fasteners.
5. Suitability for use over steel decks with open flutes.
6. Friability and handling properties and resiliency.
7. Vapor permeability.
8. Water absorption and capillarity.
9. Water adsorption (surface water).
10. Absorption of hot bitumen.
11. Resistance to chemical and solvent adhesives.
12. Melting temperatures and damage from heat.
13. Basic material and binders.
14. Resistance to freeze-thaw cycling.
15. Fire resistance, flame spread, smoke developed.
16. Density.
17. Heat-storage capacity.
18. Resistance to decay by microorganisms.
19. Food value and nesting properties for insects and vermin.
20. Probable long-term life expectancy under service conditions that will involve daily variations in exposure to moisture and temperature in a closed envelope.

Table 7.1
Amount of Insulation Required

Roof Insulation Materials

	k (per in.)	R (per in.)	Thickness Required			
			R 10		R 20	
			in.	cm	in.	cm
Fiberboard	0.36	2.78	3.6	9.14	7.2	18.29
Fibrous glass	0.27	3.70	2.7	6.86	5.4	13.72
Foamed glass	0.40	2.50	4.0	10.16	8.0	20.32
Extruded polystyrene	0.20	5.00	2.0	5.00	4.0	10.00
Foamed polyurethane	0.16	6.25	1.6	4.06	3.2	8.13*
Perlite	0.36	2.78	3.6	9.14	7.2	18.29
Glass wool batts	0.35	2.86	3.5	8.89	6.0	15.24

*Gradual replacement of foaming agents by air and water vapor can reduce the R value and increase the thickness required. Freon 11 and 12 and carbon dioxide are used. Fully aged polyurethane (3 to 6 lb per cubic foot) has a resistance of approximately 4.0 per inch, which is slightly better than fibrous glass. The absorption of vapor increases the overall dimensions.

21. Probable effect of water entering the system because of a roof or flashing leak.
22. Probable effect of moisture vapor and air entering the system because of no vapor barrier, an ineffective barrier, or because of openings at projections through the deck.
23. Potential for venting to outlets through joints between insulation slabs or through the insulation itself.

In Table 7.1, the approximate k and R values for several materials are shown, together with the thickness required to achieve a total resistance of 10.0 and 20.0.

Roof Insulation Materials

The following products can be used in roofing insulation. The materials are listed in roughly the order they appeared on the market.

Wood and cane fiberboard
Cork
Cellular glass
Glass fiber
Mineral fiber
Perlite
Polystyrene — foamed
Polystyrene — extruded
Polyurethane
Polyethylene
Polyvinyl chloride
Composite urethane and fiberglass

Thermal Insulation

Composite urethane and gypsum

Composite urethane and perlite

Phenolic

Composites of foam and fiberboard

There are about 250 manufacturers of these products in the United States. Each was trying to obtain a piece of the market on the basis of lower cost or a claim to better efficiency or because of failures in the earlier products. Distribution costs of light materials also influenced where manufacturing facilities were built. When the various materials are carefully examined it will be found they have vastly different properties and yet they are all used for the same purpose, in the same location, in a roofing system. This would seem hard to rationalize.

Residential standards for roof insulation, depending on location, utilities, and fuel cost, vary from R 20 to R 38. To achieve the R 20 level, excessive amounts of rigid insulation materials would be required, but it could be achieved with a 6-in. glass wool batt located below the roof deck in wood-frame construction. R 38 would require 13.29 in.

If one were considering placing rigid insulation either above or below a roof membrane, an R value of 10.0 might be the most practical owing to the thickness required. Below the roof, thicker insulation has the potential for greater air and vapor entrapment and causes greater movement in the exposed membrane due to thermal cycling. Above the roof thicker foam-type insulation requires positive attachment and heavier ballast.

No roof insulation, regardless of its thermal efficiency, should endanger the life of the roof membrane, which is the most important component of the system.

It has been common practice during the last 35 years to insulate roof systems by locating a low-density, rigid sheet material between the roof deck and the roof membrane. Various forms of vapor retardant materials in roll form were often, but not always, placed under the insulation to protect it from moisture vapor in the heated interior. This was due to a vapor pressure gradient, ordinary air flow due to chimney action, exfiltration due to wind forces, and ventilation rates. At the same time, building regulations have specified an air-vapor barrier on the warm side of insulated walls and a breather type of sheathing, building paper, and exterior finish on the cold side. Past and present practice described above for flat roofs, which has not proved to be free of failures, is at variance with successful practice in walls.

One can safely conclude that locating low-density air-filled insulating materials on the warm side of a vapor-tight roof covering is not logical. It is well known that vapor retardant materials located on the warm side of the thermal insulation are never absolutely vapor tight. In any event, they are never free of perforations. The chance of water penetration from above and vapor penetration from below is always high. This moisture ends up in the vapor trap and eventually destroys the roof; it does not take long. It is a case of choosing between the cost of higher heating and cooling by eliminating the insulation, and constant repairs and the replacement of the whole roofing system every few years.

It is not only the cost of roof repairs and replacement of both insulation and roofing and perhaps flashings that is involved. There are many other expenses that arise out of a condition of a structure that interferes with normal day-to-day operation.

The type of roofing failures and the time frame do not consistently or conclusively indicate poor quality conventional roofing materials or poor workmanship. They are not caused by exposure to the weather, which is not so severe that common roof coverings are unable to stand up to it for 20 years or more.

If it is not the materials or the workmanship or the weather, then what is it? The only component left is the thermal insulation introduced to reduce heat loss and to make a steel deck practical as a roofing base. If the insulation and the roof membrane are separated as in a wood frame roof (see Figure 4.2 C), or where a roof membrane is laid directly on concrete, plywood, or T&G decking, most of the roofing failure problems disappear. If it is that easy, then why don't we rethink the tight insulated roof sandwich? While it is not the time to be concerned about the welfare of the rigid insulating board manufacturers, it is the time to worry about our own buildings and our own businesses. That there are more manufacturers of insulation than roofing and a constant introduction of new insulations and new combinations could be an indication of a fundamental weakness in roofing system design. Each new insulation that comes along is reported to solve the problems of its predecessors and with a higher R value to boot. There is too much emphasis on keeping heat in and not enough on keeping water out.

When Is Insulation Needed?

Insulation in a roof system may be required for one or more of the following reasons.

1. To reduce heat loss and thereby save heating energy.
2. To reduce heat gain and thereby save cooling energy.
3. To maintain the temperature of the underside of the roof deck above the dew point of the inside air.
4. To provide a bridge over open flutes in steel decks as a base for roofing. Other materials can be used.
5. To insulate the deck from extremes in outside temperature in order to reduce thermal movement.
6. To maintain constant manufacturing or storage temperatures.

When thermal insulation is being considered for a roofing system, it is necessary to determine which, if any, of the above situations apply, and to determine if there is a satisfactory alternative. The comparative thermal efficiency of various insulation materials is not the first or most important property to examine. Not even the cost per unit of thermal resistance should be the deciding factor. Both of these properties or conditions can be quite misleading.

It is easy to think that insulation is necessary on every roof. In view of the frequency of roof and insulation failures and the increasing costs of replacement, the ultimate value of insulation below the roof membrane is questionable.

Placing insulation above the roof membrane eliminates many problems but it is

Thermal Insulation

only suitable for some buildings with sloped roof decks. The ideal insulation for the protected roof system has not yet been developed.

Some manufacturing processes produce waste heat that can be dehumidified in heat exchangers and then recycled or recirculated to warm ceilings and roofs as is done in some paper mills. There is no need to insulate to keep heat in when warm air is being exhausted to the outside. If the air is contaminated it can be cleaned up before it is recirculated.

Conventional Systems

In a conventional system the insulation protects the roof deck but not the roof membrane, which is the most important component in the system. While insulation, if it stays dry, is protecting the deck, it is helping to destroy the roof. If the insulation traps water or becomes saturated it can cause deterioration of the deck (all types).

The only sensible way to keep both deck and roof covering in the same environment, as close as possible to that of the interior, and at the same time protect both from the exterior environment, is to locate the roof membrane on the deck and then install the insulation, probably extruded polystyrene. A form of opaque ballast is required to prevent flotation of the insulation in the event of ponding, and to protect it from the sun. This is known as the "Protected Membrane System." Note that ***expanded.*** polystyrene is not suitable for this type of roof, because of its open cell structure. Opponents of the PMS object to having to slope the roof deck to drain and say it costs more because the structure needs to be stronger to carry the extra dead load. Both of these conditions should be incorporated in all roof systems but are generally left on the altar of economy. Economies such as these have contributed to serious roofing and insulation failures, roughly estimated at $500 to $600 million a year. No one has stepped forward to accept any part of the blame or any part of the horrendous cost. The chief beneficiaries are the roofing contractors (if they are not sued) and legal counselors who are delighted to earn a fee in court.

The use of roof insulation is justified to prevent heat loss from the interior of some buildings, but in southern latitudes where heat gain is more of a problem, the insulation might be omitted on some roof decks. In hot climates insulation under a roof only increases the temperature of the roof covering during the day and reduces it at night, which contributes to its demise. In any case, the amount of insulation required to reduce interior temperatures to an acceptable level would be excessive. Its principal purpose in hot climates is to prevent excessive thermal movement in the deck.

Roof coverings should be white and self-cleaning for solar heat reflection. This can be best accomplished with slightly sloped noncorrosive metal roofing over a ventilated space. Mechanical ventilation and cooling of the interior is the only reliable way to control the interior environment, e.g., temperature and humidity, in hot weather. On level roofs water ponding for cooling is dangerous and not effective. However, water spraying on sloped roofs is acceptable and effective.

Steel roof decks are covered with thermal insulation before the roof is laid, only because a roof cannot be laid over the open flutes. This very often leads to condensation of moisture in the system that damages the steel, destroys the insulation, and also destroys a flexible roof covering. Unless the insulation layer is mechanically fastened to the steel, the insulation sheets can curl up or down at the edges and the whole system can be blown away. One could say the insulation is needed to control thermal movement in the steel. This may be true, but why should the deck need all this protection when the roof is much more important? The priorities are wrong. The same amount of steel, or aluminum, can be employed as the outside weatherproofing layer and be constructed to adjust to temperature changes. It can be spray painted and kept white.

What Thermal Insulation Does to a Roof

Checking for Moisture

In the April 1982 issue of the *Roofing/Siding/Insulation Product Directory*, 15 manufacturers of various types of moisture meters or detectors are listed. These include nuclear, infrared imagery, electric resistance, capacitance, microwave, and gravimetric devices, designed to determine where water is in a roof system and how much. Their purpose is to save money for the building owner by showing what areas of the roof should be replaced and what areas can be left. One can conclude that there are enough internally wet roofing systems in the country to warrant the manufacture and use of the moisture detectors from 15 companies, and perhaps more. Their development and use came after at least a 30-year history of moisture problems in insulated roofs. They do not cure the problem; they just confirm that there is one.[1]

No attempt has been made to determine how many units are in service, or how many investigations have proved profitable to the owners of the buildings checked. The important thing is that water is present in insulated roof systems and it is only present because the insulation is there. The presence of water or water vapor eventually penetrates all the common rigid insulating materials except foamglas, reducing their thermal resistance and increasing the rate of heat flow from a heated space. The vapor pressure developed by expanding aqueous vapor can cause blisters and buckles or wrinkles in all forms of roof coverings. Water vapor travels horizontally in cork, wood fiber, and glass fiber insulation, and in the joints of plastic foams such as polystyrene, polyurethane, perlite, and phenolic materials.

The use of moisture-detecting instruments may be an excellent idea but the interpretation of readings may be difficult and inconclusive. If moisture is found or suspected, a core sample will confirm it, but how does one decide what areas of the roof should be replaced? On a large level roof, it is good practice to replace small areas here and there and attempt to make a watertight joint with what is left? Also,

---

[1]References:
Dr. R. E. Link, Jr.,"Air borne thermal infrared and nuclear meter systems for detecting roof moisture."
Wayne Tobiasson, Charles Korhonen, and Alan Van den Berg, "Hand-held infrared systems for detecting roof moisture," Proceedings of the Symposium on Roofing Technology – Sept. 1977. Sponsored by NBS and NRCA.
"Recommendations for implementing Roof Moisture Surveys in the U.S. Army," Special report 78-1, U.S. Army Cold Regions Research and Engineering Laboratory, Feb. 1978.
"Laboratory Evaluation of Nondestructive Methods to Measure Moisture in Built-up Roofing Systems." National Bureau of Standards, Jan. 1981.
National Bureau of Standards Technical Note 1146, July, 1981.

Thermal Insulation

how did the water get there in the first place, and will water penetration spread at a later date after the roof is patched?

Moisture in a roof system generally broadcasts its presence by buckles and blisters in the membrane; however, sometimes these contain dry air only and are caused by thermal movements in the membrane or insulation. A good ear can often tell the difference between wet and dry substrates by the sound during a walk over. Some wet roofs are mushy underfoot, but glass fiber insulation usually results in a mushy roof, wet or dry. Plastic foam insulations can contain water but still provide firm support for a roof membrane. In winter in cold areas water in the insulation will likely be partially frozen. A few cycles of freeze-thaw action will destroy any insulation including foamglas.

When thermal insulation under a roof is wet or soft (open cell or low density), the roof membrane can be easily punctured by point loads, such as ladders, hoists, feet, hail, or falling objects such as tools, bottles, bricks, and other construction materials.

Membranes laid directly on a solid substrate that cannot be wetted will not be easily punctured by any usual or unusual roofing activity. Flat roofs are subjected to a great deal of human traffic that has no knowledge or regard for the roof system underfoot. Unless the waterproof roof covering rests on a solid base or is protected in some way, severe and costly damage may result. There are hundreds of ways that a flat roof can be punctured, and if the substrate insulation is porous or has air spaces between the sheets, the whole system can fill up with water. There is no way it can be dried out quickly, despite claims by manufacturers of breather vents.

Try this: Place one piece of paper on your desk and try to poke holes in it with the butt end of a pen or pencil. Now place the paper on your blotting pad or typewriter pad and see how easy it is to puncture. This is not a scientific test but you will get the point.

Whenever water replaces air, insulating values drop significantly because the thermal conductivity of water is approximately 20 times that of air (in confined spaces). The conductivity of ice is 80 times that of air. If insulation in a roof system becomes wet and the thermal conductivity increases by 20 times, then it ceases to be a useful component in the system. If the water in the insulation freezes, which it can do under certain circumstances, the rate of heat loss might be unbelievable. A heated building might, however, lose enough heat to keep ice from forming in the insulation. Water at 32°F in the insulation would normally move the dew point down to the vapor barrier causing interior water vapor to condense on the underside of the deck if it had low resistance to heat transfer, and result in water dripping back into the building just as if the roof was leaking.

Example

Consider the case of a bakery building in Edmonton, Alberta.

A single-story brick building a city block long was designed and used for a large automated bakery of bread and cakes. The roof system consisted of a level steel deck, two plies of tarred felt vapor barrier mopped in hot pitch, two inches of cork insulation in one layer, and a four-ply pitch and gravel roof. There were an extraordinary number of penetrations of the deck and roof for vents, skylights, hatches, and the like, and a reasonable number of roof drains that did not drain the roof very well. In the winter, they did not drain at all.

A few years after the bakery commenced operations the roof dripped water over the whole area. In the summer when the roof was examined no flaws could be found. Several cut-outs were taken and all showed a well-laid roof, but it was discovered that the cork insulation was either wet or completely disintegrated into mush. For business reasons and perhaps also because the leaking could not be tolerated, the bakery was moved to another city. After the building was vacated an examination in the winter showed six inches of snow, two inches of ice, and two inches of water on the roof. Water was still dripping into the empty building over its entire area. Moderate heating was being maintained to keep the building alive. Holes drilled at random in the steel flutes at center span released approximately half a bucket of water from each hole.

The leaking problem was never completely resolved. The undesirable features of the roofing system were

1. A tarred felt vapor barrier.
2. A level roof deck of steel.
3. No expansion joints in deck or roof.
4. Too many roof penetrations.
5. Pitch and gravel in an area where the mean annual temperature difference was $+90°F$ down to $-45°F$ ($135°$)
6. Only two inches of cork when at least four inches or more would have been appropriate. Taking k at 0.27 per inch (1959 Guide), the U value of the system is 0.12 and the R value is 8.58. The winter design temperature, 5% basis is $-30°F$ and 10% basis is $-25°F$. A thermal resistance value of 8.58 is not high enough considering the exterior winter temperatures, and the moderate to high inside relative humidity.

Water penetration into plastic foam-insulated roof systems may be difficult to detect in the early stages by nuclear or infrared means because the water is concentrated principally in the joints between the sheets. However, there could be some water between the insulation and the deck or vapor barrier. General or overall penetration gradually spreads from the top surface down.

Wet insulation causes loss of bond to the roof membrane which is then free to move, i.e. shrink or expand due to thermal cycling. This can result in cracking, buckling, damage at building perimeter, and lifting of the roof membrane by wind suction. It is important to remember that condensation of moisture vapor in an insulated system occurs first at the roof membrane/insulation interface. This wets the upper surface of the insulation, and if there is no mechanical bond the adhesion between the membrane and the insulation is lost.

A steel roof deck with a relatively moisture-resistant insulation material, mechanically fastened directly to the steel with no vapor barrier, will very quickly develop problems if the building is heated or air conditioned, or if it is a high-rise structure.

High-rise buildings of steel or concrete are vulnerable to air and moisture flow into insulated roof systems through penetrations in the roof deck, even if the insulation is protected with an air/vapor barrier or seal. The chimney-stack effect in high-rise

Thermal Insulation

buildings is difficult to overcome and is the reason why the upper floors are dangerous in a fire.

The interior air pressure in buildings is often maintained slightly above atmospheric pressure in order to restrict or prevent the inflow of outside air. If this imbalance is not carefully controlled to under $1/2$ inch of water, the interior heated air tends to penetrate the building envelope, including the roof system, until it reaches a cold surface where it will condense. If the dew point happens to be below freezing, the formation of frost or ice is possible, and in certain cases can do considerable damage to roofs, curtain walls, and masonry surfaces by spalling. The insulation in a roof system is less likely to be wetted if the building can be operated at a negative pressure instead of a positive one. This has been proved in paper mills with wood roof decks covered with a mopped asphalt felt vapor barrier, wood fiber insulation, and an asphalt built-up roof. Paper mills operate under very high air temperatures and relative humidities, although new machinery designs have improved the machine room conditions immeasurably.

Roof Membrane Cracking and Buckling

Insulation above the deck does not provide a continuous stable surface for a roof membrane. Insulation sheets or panels 2 ft $\times$ 4 ft have 75 lineal ft of crack or joint in one square of roof (100 ft^2). Sheets 3 ft $\times$ 4 ft have 58 lineal ft. The number of joints across a 100 ft square roof (100 squares) are 50 in one direction and 25 in the other direction.

In the days when cork insulation was popular (in other words, cheap) the slabs were only 12 in. $\times$ 24 in. This meant that crack length on one square of roof was 150 ft. In the 100 square roof mentioned above, there would be 50 joints in one direction and 100 in the other. Cork was not always well anchored because of its extreme porosity and varying thickness, especially with coal-tar pitch. To obtain a reasonably smooth even surface it had to be laid in multiple layers. The amount of asphalt and pitch absorbed by the open-textured cork was considerable, which in turn reduced the thermal efficiency ($R - 3.0$ to 3.7 per inch).

If there is firm attachment of the roof membrane to insulation by hot mopping, any expansion or contraction of the insulation will be reflected in the membrane in the form of buckles or splits. Any shrinkage in the roof membrane caused by low temperatures will pull the individual pieces of insulation together, closing up the joints. When this occurs, roofs pull away from perimeters toward the center and sometimes dislodge eaves' construction materials.

To restrict movement, the insulation should be mopped firmly to the deck or suitable vapor retardant, which in turn must be mopped solid to the deck. The term "vapor retardant" is now generally preferred to "vapor barrier" because it is admitted most membranes are not perfect. Even if they were, both air and vapor can find avenues of escape through perforations in the roof deck and around the vapor retardant membrane. In the three apartment buildings described in chapter 16, the poured concrete decks would permit vapor movement at a very slow rate and yet air and moisture moved into the polystyrene insulation at the deck openings and between the insulation and the deck. On building 1 there were 60 such openings in an area of 8,100 ft^2. This building had three roofs in ten years. Foam insulation sheets are rigid and are not softened by normal roof temperatures; therefore they rarely follow the irregularities in the decks, and permit air movement between the insulation and the deck. Foam insulations cannot be laid in liberal moppings of hot asphalt so that all air spaces are filled. This could be one of the worst features of the material in roofing work.

The shrinkage and expansion of roof membranes in the cross machine direction (CM or CMD), is approximately double that in the machine direction (MD). This applies to tarred and asphalt felt made on a single-cylinder paper machine. Therefore, if the number of joints in each direction is taken into account and felt is laid parallel with the long dimension of the insulation sheets, the potential for movement is four times as great across the machine direction as in the machine direction. However, if the felts are laid across the width of the insulation, the potential for movement in both directions is equal. This also applies to most glass matt felts which are laid up in a jack-straw pattern with no predominant grain or fiber orientation. Unfortunately, nearly all roofs are laid with the long dimension of insulation sheets parallel with the machine direction of the felt. This is the easiest way for a roofer to lay the insulation and roofing together, but not the best way due to the 4:1 ratio for shrinkage and expansion movements. This could be one of the principal reasons why roofs split and buckle (see Figure 7.1).

There are so many factors that affect tensile strength that it is difficult to make any conclusive comparisons between all the different felts available. These include the type of saturant, temperature, rate of applied stress, magnitude of strain before rupture, and characteristics of the mopping bitumen.

In "Fundamentals of Roof Design" (CBD 67, July 1965), G.K. Garden states: "The extreme temperature variations between night and day in a black roof material over insulation can exceed 140°F and the seasonal variation may be over 250°F.[2] Atmospheric pollutants contacting roof materials, especially when combined with water, can accelerate deterioration."

If roofs were free and clear of any obstructions and were of a size within the bounds of the membrane's ability to withstand temperature stress, things might not be

---

[2]Northern states and Canada. In southern and Pacific Coast states the daily and annual variations would be less.

Table 7.2*
Coefficients of Thermal Expansion for Roof Membranes

Type of membrane	Coefficient per deg F $\times 10^{-6}$			
	30 to 0 F		0 to −30 F	
	L	T	L	T
Organic felt & coal-tar pitch	22.3	36.0	19.3	29.5
Organic felt & asphalt	2.7	12.6	13.9	37.5
Asbestos felt & asphalt	4.8	18.1	19.5	37.5
Glass felt & asphalt Type 1.	8.9	10.1	35.1	46.4

L - machine direction   T - cross machine direction

*From CBD 182 Nov. 1976. Originally Mathey, R.G. and Cullen, W.C. "Preliminary Performance Criteria for Bituminous Membrane Roofing". National Bureau of Standards, Bldg. Sci. Ser. 55, 1974, 19p.

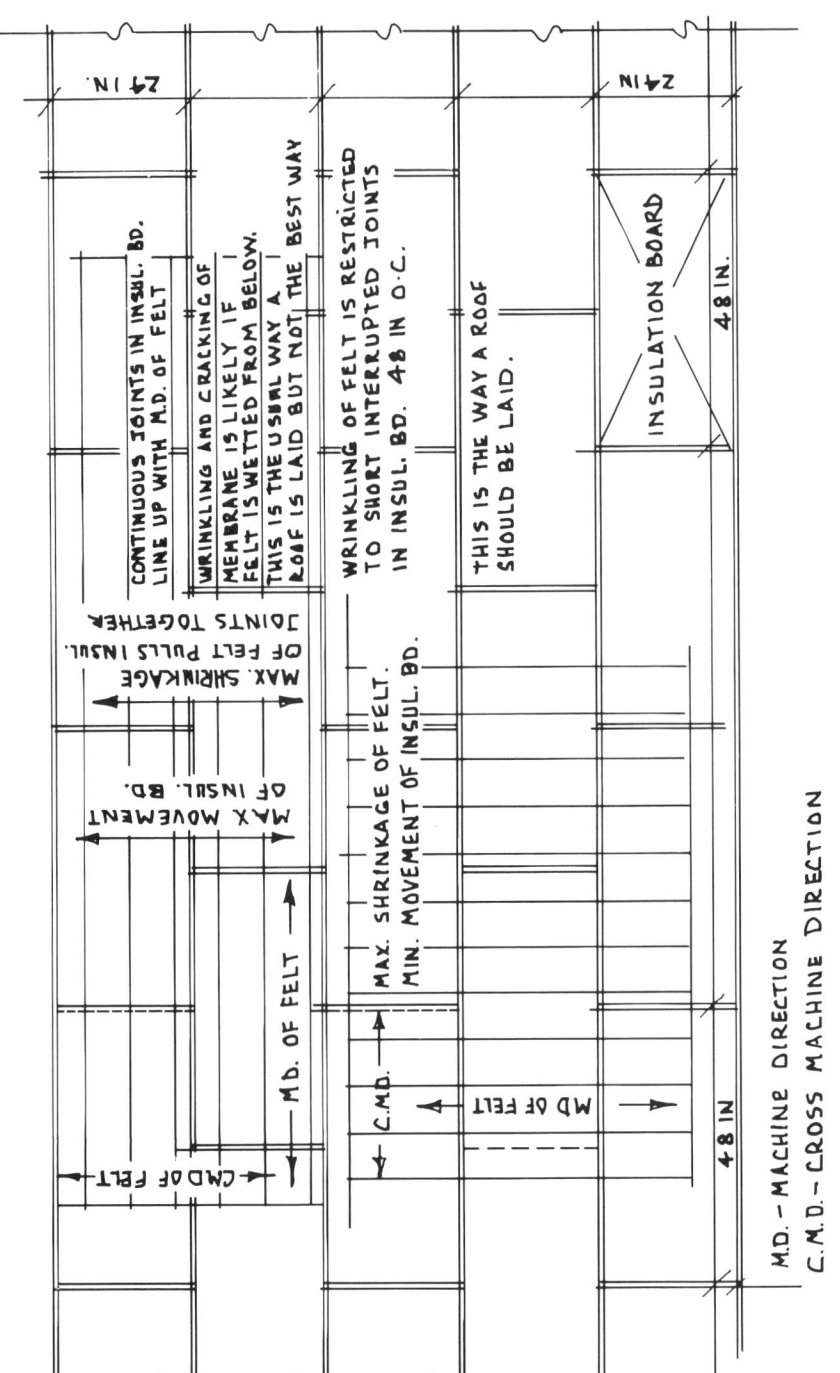

Figure 7.1 Roofing felt directions over insulation.

too bad. However, roof areas are large and are interrupted by ventilators, hatches, penthouses, skylights, and other obstructions. Any tendency for a roof to split may be triggered at one of these obstructions, which interfere with or interrupt the continuity of the membrane in a haphazard manner.

Some 20 years ago it was noticed that poor quality glass felt roofs split in long sweeping curves, presumably following the weakest areas of the glass matt and thin moppings. The curious thing was that the cracks always started at an obstruction in the roof. The insulation under the roof was glass fiber of about 8-lb density with a paper covering on one side. No satisfactory scientific explanation was ever given for these failures. The glass matt was manufactured in Canada based on German practice. Improvements have been made to glass fiber roofing felts or matts since 1960.

Felt Laying Patterns

The long-standing procedure of laying felt shingle fashion may have contributed inadvertently to buckling and splitting. In Figure 7.2 felts are shown laid two ply, three ply, and four ply, with the standard 2-in. head lap. This drawing shows the membrane always has one more ply of felt and one more mop of bitumen at the head lap, assuming the felt plies are laid exactly right. This means the system is greater in tensile strength in the MD at the head laps than it is in between. It is also more likely to buckle between the head lap in the CMD because of the fewer plies of felt. In other words, the membrane is not consistent in the number of plies or tensile strength across the whole area.

If the roofer is careless or if the felt does not roll out straight along the lines provided on the felt, the head lap may be increased in one course and reduced in another. The latter case is illustrated between the two-, three-, and four-ply section. It will be seen that in the two-ply roof there could be only one ply of felt instead of three. In the three-ply roof there are only two plies instead of four, and in the four-ply roof there are only three plies instead of five.

This situation could be responsible for splitting, especially in the two- and three-ply examples, and could be part of the reason for the failure of many two-ply roofs laid in the 1970s with coated felt.

Research on Roofing Splits

The following is an excerpt from "Built-Up Roofing Tension Splits," by G. Norman Mosely, Product Manager, Fiber Products, Domtar Construction Materials Ltd., Montreal, Quebec. It was first published in *Building Research*, November–December, 1964.

> In the case of a membrane applied directly to a roof deck, the problem is simply a matter of bonding one to the other. In such construction, contraction splits are unlikely. In the case of a membrane over insulation over a vapor barrier, not only the lamination between the various components, but the horizontal shear strength of the insulation, particularly at its surface planes is involved. A further consideration is the ease with which the insulation might compact at the joints or within its

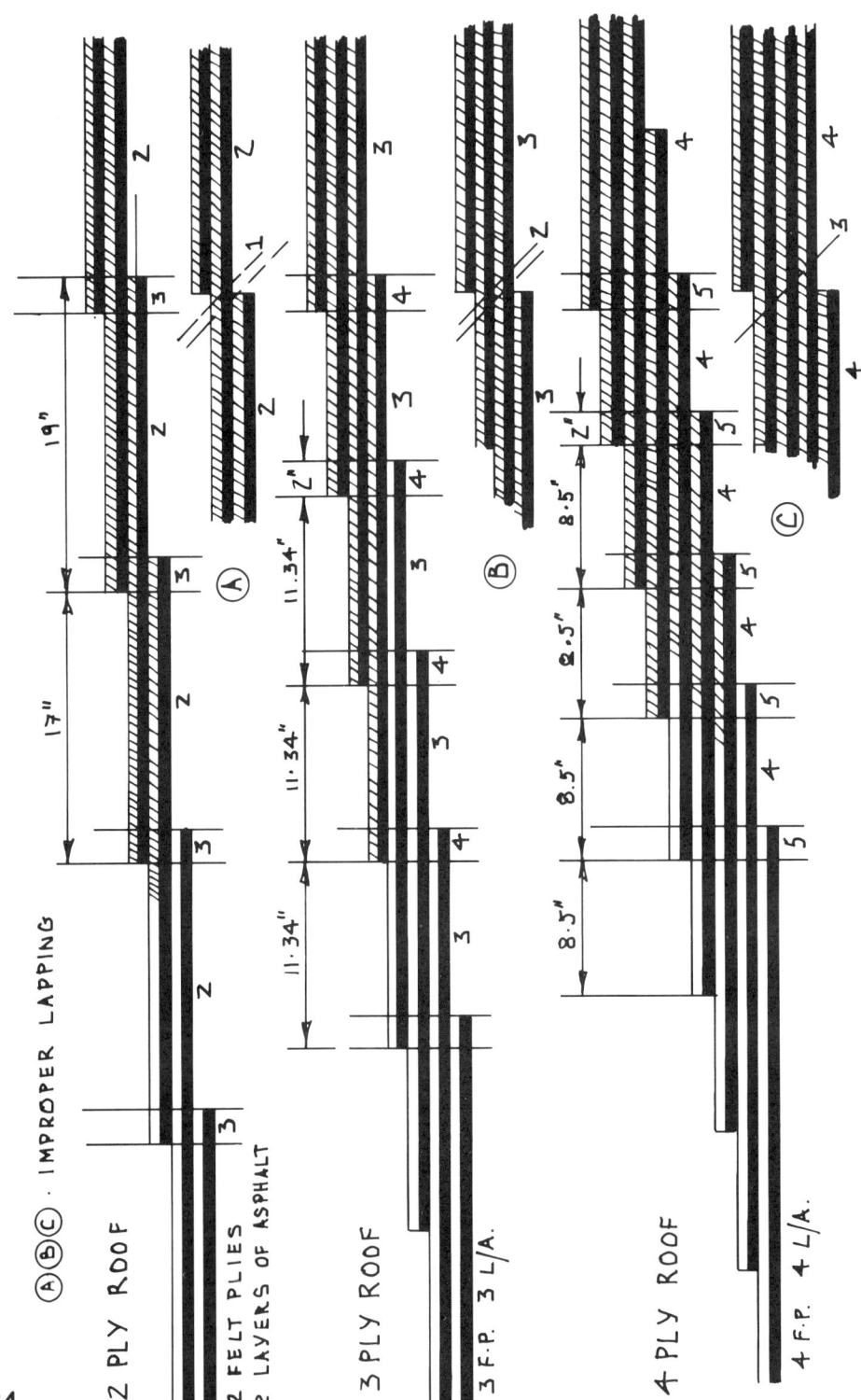

Figure 7.2 Shingle mopped felt (top surfacing ommitted).

mass. Because the ductility of most laminates will be greater at high temperatures than at low, and because the warmest plane in the assembly during cold weather is that nearest the deck, it is here, disregarding other factors, that slippage is most likely to occur.

And again in "Properties of Bituminous Membranes" (CBD 74, February 1966, DBR/NRC Canada), G.K. Garden writes:

> Splitting occurs in a bituminous membrane when its ability to accommodate strain is exceeded. The stress producing strain may be caused by temperature variations, shrinkage due to aging, moisture changes, and the behaviour of the structure to which it adheres. Adhesion to a rigid substrate can cause the strain to take place uniformly over the membrane, but any differential movement in the substrate at the joints or cracks causes great risk of high local strains and splitting.

The following is an excerpt from a 1981 talk on an overview of roofing and new roofing materials (not published), by Maxwell C. Baker, Senior Officer, Division of Building Research, National Research Council of Canada.

> The demand for controlled climates in many buildings for processes and for human comfort has brought about two changes — the almost universal use of insulation, and a greatly increased use of air conditioning and humidification. The effect of the insulation on roofing was not always appreciated when it was first introduced. The insulation subjects the exterior surfaces (the roofing membrane) to higher temperatures in the summer and daytime, and colder temperatures in the winter and nighttime, and a wider range of temperatures than is the case for an uninsulated roof. The humidification creates larger vapor pressure gradients across the building enclosure, and provides moisture that can be driven into walls and roof by air pressure differences as well as by vapor pressure differences.
>
> The changes in building design and practice are fairly obvious, but they have implications that may not be so obvious. The physical nature of the insulation and the roofing membranes, all affect the performance of the roofing. The structural elements produce deflections, differential deflections, and a variety of movements that can affect drainage and the integrity of materials attached to them. The early use of above-deck insulation consisted of fiberboard and cork, in board form. They were mopped solid to decks with bitumen, and roofing membranes were mopped solid to them. Such attachment or anchorage by adhesion was close to 100% effective. The need for more efficient thermal insulation, however, has led to the development of a variety of insulations, with vastly different properties, that are still used in essentially the same manner as the earlier insulations. The large number of roof deck and insulations available make the number of combinations that are possible very great indeed, and it seems quite likely that the interaction between

Thermal Insulation

combinations may be somewhat different. In recent years, it has also been common to combine more than one type of insulation on the same roof in factory-prepared composites, or on-the-job composites. Some of the insulations available are not entirely compatible with bituminous materials, and the usual practices, for the application of the earlier materials, do not necessarily produce the same attachment with the newer materials. The implications of poor or nonexistent attachment between components, and differing rates of movement between them, is still not fully appreciated or assessed.

A Frank Statement About Roof Insulation

After careful analysis of the continuing expensive problems with thermal insulation tied to flexible roof coverings, the author is convinced they should be divorced from each other whenever possible. Roofing is a high-density vapor-tight membrane designed to shed water, while insulation is a low-density air filled material designed to reduce heat flow. An air/vapor barrier is introduced to stop movement of air and vapor into the insulation and roofing from the heated interior. It is obvious that these three functions are very different and must be employed separately, not as a homogeneous assembly in the space of an inch or two. In this situation, when one component fails the whole fails. This is one of the reasons why reroofing a building is so expensive. All three components in the system—insulation, vapor barrier, and roof—usually have to be replaced when it may only be the roof that is faulty or, more likely, the insulation is first saturated with water. There is little doubt that an ineffective air/vapor barrier or none at all can cause the destruction of the insulation or fill the system with water. In this case, the roof covering is also destroyed.

When moisture is admitted to a roofing system from whatever source, an increase in temperature can produce pressures exceeding 850 psf, that can buckle and blister the thermoplastic roof coverings. All built-up or single-ply membranes are thermoplastic. It is evident that the only time a thermoplastic roof covering can be safely used is when there is no capacity for the accumulation of moisture or air under it.

It has been estimated that a decrease of 0.1 in the U factor of the nation's low-sloped roofs would save nearly 20 million barrels of crude oil, or approximately 116×10^{12} Btu annually.[3] The emphasis is on increasing thermal resistance without due regard for the safety of the roof membrane.

For many buildings there is a good case for a secondary rigid deck above the insulation to support a flexible roof covering or a rigid metal roof, separate from the insulation layer, and not affected by water vapor or air pressure. It would need to be designed to adjust to thermal expansion and contraction and be noncorrosive. Such roofing materials and systems are available now in zinc-coated or bitumen-coated steel, terne-plated steel, enameled steel, and various forms of aluminum and copper. In such a system the thickness and thermal resistance of the insulation could be much

---

[3] National Roofing Contractors Association, "Good Roofs Save Energy," *The Roofing Spec.*, July, 1976, pp. 13–15.

greater than is now possible or safe with the thin insulated sandwich. This represents a continuing and long lasting savings in energy cost at no risk to the roof.

While a metal roof might cost more initially than a flexible thermoplastic covering, it could be designed to last perhaps three or four times as long and could be repaired or replaced without having to interfere with the other components in the system. A small incline would permit water flushing to remove air contaminants and at the same time cool the roof. As for appearance, there is no comparison between a white ribbed metal or asbestos cement roof and a dirty black graveled one. White marble or limestone chips on built-up roofs do not stay white very long in most areas.

A great deal of money is being wasted on roof repairs, coating materials, investigations for moisture, increased use of energy because of wet insulation, downtime, and new roofs laid over old on existing badly designed buildings. Generally, roof investigations and repairs rarely dig deep enough into the basic cause of roof failures. These expenses never stop. They keep increasing year by year.

Reflective Surfacing

Provision of a reflective surface is the answer to the problem of high peak temperatures in the membrane, as well as being part of the answer to the problem of keeping the structure cool. Reflectivity of different surfaces is defined as the numerical ratio of the radiant energy impinging on the surface. The coefficients of solar reflectivity of different surfaces have been determined and some approximate values are listed in Table 7-3.

Table 7.3

Surface	Coefficient of Solar Reflectivity*
Mastic asphalt and bitumen compounds	0.07
Bituminous roofing felt, dusted	0.10
Asbestos cement roofing, old and dirty	0.17
Concrete tiles (color not specified)	0.35
Asbestos cement, new	0.39
Asbestos cement, freshly washed	0.60
Galvanized iron, old and dirty	0.10
Galvanized iron, new	0.35
Copper, tarnished	0.36
Copper foil, new	0.75
Aluminum foil, new	0.87
Slate, blue gray	0.15
Marble, white	0.56
Sand, fine and white	0.59
Paint, light gray	0.25
Paint, red (shade not specified)	0.26
Paint, aluminum (varies with type of paint)	0.46
Paint, light green	0.50
Paint, light cream	0.65
Paint, white	0.75
Whitewash	0.80

*Reflectivity values are from K.G. Martin, C.S.I.R.O. Melbourne (Commonwealth Scientific Industrial Research Organization).

Thermal
Insulation

Suggested References

Cullen, William C. "Effects of Thermal Shrinkage on Built-up Roofing." U.S. Department of Commerce, National Bureau of Standards. NBS Monograph 89, March 4, 1965.

Solvason, K.R., and Handegord, G.O. "Ridging, Shrinkage and Splitting of Built-up Roofing Membranes." Building Research Note 112, June 1976, National Research Council Canada, Division of Building Research.

8

Air-Vapor Barriers

In order to control the movement of moisture vapor by diffusion from warm interior spaces to the colder outer spaces in a structure, a vapor-impervious membrane is installed on the warm side of the wall, ceiling, roof, or floor. While this can be effective in preventing condensation due to cooling, it depends to a great extent on the air tightness of the membrane. In an insulated flat-roof system, it is imperative that there be no direct air paths from the interior into the insulated sandwich. Providing some means of venting the sandwich often contributes to the flow of air and vapor from the interior, which may result in condensation somewhere in the system. It can be argued that the elimination of an air-vapor barrier will allow a reverse flow of moisture vapor when conditions are favorable (i.e., during summer). This is a dangerous theory because of the absorptive nature of most insulation materials and because of the rapid changes in vapor pressures daily and throughout the year. It would be difficult to establish design criteria owing to geographical variations and variable design requirements for interior environments.

Some roofs laid on organic insulation over steel decks without a vapor barrier have been completely destroyed in the first or second year by the absorption of construction-phase moisture during winter construction.

9

Ventilation of Roof System

Venting has already been mentioned in chapter 7 because insulation, vapor barriers, and ventilation are so closely related. In about 1940, when built-up roofs were first insulated, the materials used were wood and cane fiberboard with a density of about 17 lb per cubic foot and cork with a density of about 12 lb per cubic foot. Little was known at that time about the need for vapor barriers. Even in cold climates not more than 2-in. thicknesses were used, and the thought of ventilating the system was not considered. Wood or concrete were the common deck materials. Since 1940 the types of decks, roofing, and insulating materials have proliferated to the point where endless combinations are possible. All are plagued by moisture accumulating in the system from the building interior, because of roof or flashing leaks, or because moisture was built into the system during construction.

Moisture becomes apparent through the appearance of blisters, buckles, and splits in an otherwise good roof. It may be soft and spongy or may be floating on a pool of water. There may be direct leaks through the deck into the building. Hidden moisture is detected by inspecting cutouts and through expensive procedures involving infrared photography and electronic roof scanners. Even when it is evident that there is water in the system, it is not easy to determine its extent or quantity and, most important of all, how it got there. There is no point in modifying the system by installing some sort of internal air circulation for drying unless the last question is

Ventilation of Roof System

answered. In a discussion to determine responsibility, the buck is passed between the architect, general contractor, roofing contractor, deck manufacturer and installer, insulation manufacturer, and roofing materials manufacturer. Very often the question is settled in civil court, with the wrong party to the action sometimes being accused and convicted. The result of course is not a jail sentence or a fine, but a court order to replace the system in its entirety. The cost can bankrupt a roofer or general contractor and force an architect to carry expensive liability insurance.

On new roof systems it has been suggested that insulation sheets be deliberately channeled to allow free movement of air and water next to the vapor barrier. In addition, perimeter flashings are designed to vent the system, and various forms of venting devices are specified for the body of the roof – one for each ten squares – an arbitrary decision. It would appear that this sort of thinking is based on the belief that the entry of water is inevitable and that the ventilation, if indeed there is any, plus a sloped deck for drainage at the vapor barrier level will keep the system dry.

On old roofs that are obviously wet under the roof membrane, breather vents are suggested. Since it is generally impossible to rebuild the perimeter to allow fresh air to enter the system, the value of the breather vents is questionable. They might even contribute to the flow of vapor from the building into the roofing system if there is a flaw in the vapor barrier.

If a defect in the roof covering or in the flashings has not been found and corrected, venting is a waste of time. If the air spaces in the insulating layer and perhaps the insulation itself are saturated, there is no chance for air to move toward the exits in any sort of drying action. Even under ideal conditions the natural forces required to move moisture vapor in a horizontal direction are very small, being generated by wind or by heating and cooling, which creates a pumping action. Outward movement of aqueous vapor in a roof system requires that the vapor pressure inside exceed that outside. Reports of the effectiveness of venting are far from conclusive one way or another; therefore, it would seem advisable to construct the system so that the potential for moisture entrapment below the membrane does not exist.

10

Roof System Specifications

The State of the Art (1979–1982)

The area of built-up roofing constructed each year in the United States covers approximately 3 billion ft^2 or 108 mi^2. A conservative estimate of the total amount of the nation's existing low-slope roofing is 25 billion ft^2 or 900 mi^2. While built-up roofing in general performs satisfactorily, premature failures cause inordinate expenses for owners and unneeded complications for roofing contractors, general contractors, architects and engineers. Roofing contractors and roofing manufacturers indicate that a probable failure rate of 4 to 5 percent may be accurate; however others quote even higher figures. Many of these problems are attributable to moisture on one or more components of the roof system.[1,2]

If a 4% failure rate is accepted, the number of squares (100 ft^2) of flat roofing in trouble because of some fault is 1.2 million. It is too simple to suggest that the total cost is 1.2 million times $400 or $500 per square, or $480 to $600 million. It is possible, however, that because of the degree of expensive scientific investigation of roofing problems being undertaken there must be something wrong. This investiga-

---

[1] Based on a report by Herbert Busching, Robert Mathey, Walter Rossiter, Jr., and William Cullen. R/S/I April 1979, "Effects of Moisture in Built-up Roofing."

[2] The author takes no responsibility for the accuracy of these figures.

Roof System Specifications

tion has been going on for at least 30 years and yet the United States is still suffering a considerable early failure rate at an ever increasing cost per square of roof. On some types of buildings the failure rate is much higher than 4%.

It is obvious that the problems have not been solved with supposedly tried and true asphalt and tar roofing materials and specifications, and yet a new breed of elastomeric and rubber sheet materials and modified asphalt coatings are being introduced into the market; some say as much as 15 percent now and 25 percent by 1985. We have yet to find out the problems they will add to the ones we already have. In addition to the need for a watertight roof system, there remains the basic requirement for a system to provide a satisfactory degree of environmental separation. Changing the form of the waterproof covering will not change this.

Roofing Research

There has been a phenomenal amount of research by government, university, and other nonprofit organizations as well as by roofing associations and manufacturers of materials. The work that has been going on for many years in Australia, Canada, England, France, Germany, Japan, Sweden, and the United States is far in excess of the basic simple purpose of a roof, which is to protect a building from external moisture and solar energy. Reasonably clear but sometimes controversial guidelines for better roofing have resulted from all this learned activity. Following is a partial list of the published researchers, principally in the United States and Canada, who have been engaged in this work. It is still continuing. The omission of many helpful researchers is regretted and is not intentional.

Abraham, Herbert — Asphalt and Allied Substances, 1962
Alexander, S.H. — Kenneth Tator Associates
American Society of Testing and Materials
Anderson, L.O. — USDA Forest Service, Madison, Wis.
Appleton, William H. — NBS
Asphalt Roofing Industry Bureau, Washington, D.C.
Baker, Maxwell, C. — DBR/NRC
Ball, Walter — DBR/NRC
Blaga, A. — DBR/NRC
Boone, Thomas H. — NBS
Brotherson, Donald E. — University of Illinois
Busching, Herbert
Canadian Roofing Contractors Association, Ottawa, Ont.
Craig, Willis G. — NERSICA, February 1954
Cullen, William C. — NBS
Dalgliesh, W.A. — DBR/NRC
Davis, D.A. — ASTM STP 603, 1976
Dickens, H.B. — DBR/NRC

Fricklas, Richard L. — R.I.E.I
Garden, G.K. — DBR/NRC
Gibbons, E.V. — DBR/NRC
Giles, L.W., Jr. — Navy
Granum, R.M. — University of Minnesota
Griffin, C.W. — BUR Systems — McGraw-Hill
Gumpertz, W.L. — Simpson Gumpertz and Heger Inc.
Hamada, H.
Handegord, G.O. — DBR/NRC
Hansen, A.T. — Canadian Builder, 1963
Harmathy, T.Z. — DBR/NRC
Hedlin, Charles P. — DBR/NRC
Hoiberg, Arnold J. — NBS
Hutcheon, N.B. — DBR/NRC
Jacoby, M.E. — Owens Corning Fiberglas
Jenkins, David R. — NBS
Jenkins, J.H.
Jones, P.M. — DBR/NRC
Joy, Frank A. — Pennsylvania State University
Kariya, H.
Keuster, George H. — DBR/NRC
Kishitana, K.
Knab, Lawrence I. — NBS
Koike, Michio — B.R.I., Tokyo
Krenick, M.P. — ASTM STP 603, 1976
La Cosse, R.A.
Lally, H.O. — DBR/NRC
Long, E.G. — Johns-Manville Research
Louden, A.G.
Lund, C.E. — University of Minnesota
March, F.O. — P.I.S. Roofs & Roofing, Brighton, 1974
Martin, K.G. — C.S.I.R.O., Melbourne, Australia
Masters, D.J. — Building, 1967
Mathey, Robert G. — NBS
McCormick, E.J. — Roofing Consultant
McInnes, H.W. — Building, 1967
Mertz, Edwin — N.R.C.A. Chicago
Midwest Roofing Contractors Association

Roof System Specifications

Mirra, Edward — R/S/I, December, 1976
Mosely, G.N. — Domtar Construction Materials Ltd., Montreal P.Q.
National Roofing Contractors Association, Chicago
Oliensis, G.L. — Director of Research, Lloyd Fry Roofing Co., Summit, Ill.
Pierce, E. Thomas — NBS
Poirson, A.
Powell, Frank J. — NBS
Robinson, Henry E. — NBS
Roofing Industry Educational Institute — R.I.E.I., Denver, Colo.
Rossiter, Walter E. — NBS
Rush, Richard — Progressive Architecture, September 1978
Schreiber, E.T. — General Services Administration, Washington, D.C.
Schriever, W.R. — DBR/NRC
Shuman, E.C. — Pennsylvania State University
Siu, MCI
Skoda, Leopold F. — NBS
Solvason, K.R. — DBR/NRC
Stafford, B.F. — DBR/NRC
Stafford, Robert M. — Roofing Consultant, NC
Stephenson, D.B. — DBR/NRC
Tamura, George T. — DBR/NRC
Tator, Kenneth — Associates, Coraopolis, Pa.
Thomas, J.
Tibbetts, D.C. — DBR/NRC
Tobiasson, Richard J. — NBS
Tobiasson, W.N. — CRREL (Cold Regions Research and Engineering Laboratory – Corps of Engineers)
Turenne, R.G. — DBR/NRC
Tye, R.P.
Warden, Warren B. — NAS/NRC Washington
Warden, F.C.
Wilson, A.G. — DBR/NRC
Wilson, F.C. — Architectural Record, October 1962
University of Illinois, Small Homes Council, Building Research Council
University of Wisconsin, Department of Engineering and Applied Science

Contributors to Roofing Problems

The obvious conclusion is that there is something wrong with what is called "the state of the art," which has evolved through input from about 24 sources or contributors.

Reasons for Roofing Problems

These can be tabulated as follows:

1. Architects and engineers
2. Builders and miscellaneous subtrades
3. Roofing and sheet-metal contractors
4. Roofing materials manufacturers
5. Bitumen manufacturers
6. Insulation manufacturers
7. Roof deck manufacturers
8. Mechanical fastener manufacturers
9. Special adhesive manufacturers
10. Sheet-metal manufacturers
11. Roof drain manufacturers
12. Manufacturer's application or "use" specifications
13. Regional and national roofing associations' specifications
14. Independent specification writers
15. Roofing inspectors
16. Local ordinances and building codes
17. Local fire regulations
18. Government and armed services standards for materials
19. American Society of Testing and Materials (ASTM)
20. Canadian Standards Association (CSA)
21. Underwriters Laboratories, Inc.
22. Factory Mutual Research Corporation
23. Research organizations — government and university
24. The building owner

It is small wonder that, with so many interested parties involved to varying degrees, there is confusion in the industry with specifications, and less than perfect performance of roofs. It is obvious that there are too many specifications and too many materials to do a comparatively simple job of waterproofing a building. One cannot attach any blame to one or even a special group for the problems that are encountered, but it is possible from the hundreds of field investigations to consider the following reasons for roofing problems prepared and published in 1979.[1]

Reasons for Roofing Problems

1. Dead-level roofs that pond with water.
2. Moisture-sensitive and absorptive organic roofing felt and insulation materials where repelling water is the primary objective.

---

[1] John A. Watson, *Roofing Systems: Materials and Application*, Reston: Reston Publishing Company, Inc., 1979.

Roof System Specifications

3. Low-density plastic foam and glass-fiber insulation.
4. All forms of thermal insulation sandwiched between a vapor barrier and a built-up vapor-impervious roof covering.
5. Insufficient number of plies of roofing felt and bitumen to resist thermal shrinkage stresses.
6. Roof decks containing excessive moisture, causing blisters.
7. Roof decks that allow excessive movement between preformed units.
8. Roof decks that allow excessive temporary or permanent deflection under load.
9. Unpredictable shear strength at the roofing-substrate interface.
10. Any form of fixed immovable mechanical fastening of soft compressible insulation to roof decks, or nailing roofing components through the insulation.
11. Use of mineral-surfaced cap sheet roofing over mopped felts, especially when nailed on low slopes.
12. Improperly designed flashings.
13. Roof drains of insufficient size, number, style, and placement to drain a roof quickly.
14. Lack of regular maintenance by owner.
15. Mixing coal-tar and asphalt products in the same roof.
16. Mixing asphalt saturants, filled coatings, and unfilled hot moppings that are not compatible with each other.
17. Use of roofing gravel that is not opaque to ultraviolet light.
18. Application of membrane roofing on roof inclines that are too steep or that have variable inclines.
19. Roof specification not suitable for the interior or exterior environment and roofing in wet weather.
20. Use of black smooth-surfaced asphalt roofs over insulation and without a heat reflecting surface coating.
21. Installation of electric conduit on top of the roof deck and within the thermal insulation layer.
22. Lack of control over construction-phase moisture within the building.
23. Poor-quality application procedures and workmanship, and also materials handling and storage.
24. The roofing bond.
25. Damage by other trades.

In addition to the reasons enumerated for roofing problems, a number of problems also exist in the roofing systems themselves. These include

1. An uncommon lack of interest in durability or long service life.
2. Increase after World War II in the area of large one-story factories and warehouses.

3. Slopes in large roofs have become even more difficult to build; therefore the roofs do not drain readily.

4. Basic building principles are ignored in the interest of simplicity of design.

5. Wood and concrete roof decks are replaced by precast or premolded units for easy handling. The relationship to the roofing is often ignored.

6. Movements and deflections in new deck units are considerably different, one from another.

7. Controlled interior environment creates harmful environmental gradients.

8. Air leakage from high-rise buildings results in damage to roofs.

9. Roofing operation is not protected in winter like other building trades.

10. Too many unpredictable insulating materials; attachment and compatibility are questionable.

11. Mechanized application does not always make a good roof: Equipment faults are not noticed and corrected; foaming and skips of bitumen can be hidden.

12. High wheel loads can damage decks, vapor barriers, insulation, and roofing.

13. More, as well as permanent, mechanical equipment is mounted on flat roofs. Flashings become more complicated, and servicing of machinery often damages roofs by liquid spills or simple abuse.

14. Scientific research does not always deal with problems relative to roofing as a whole entity. Roofing consultants are not as knowledgeable as they think they are.

15. As a result of a multitude of problems with BUR since 1940, the chemical industries in Europe and the United States began selling experimental materials. Many failed in an attempt to solve problems with yet another cure-all material.

16. Th expected 20-year service life is often considered reasonable when it is much less than the life expectancy of a building. Replacement costs are now 10 to 20 times the cost of the roof 20 years ago due to increased costs of materials and labor. A service life of only 20 years for a flat roof is not realistic when a relatively inexpensive sloped shingle roof will often last that long or longer.

17. One thin layer (1 mm) of plastic or rubber is not likely to cure the flat roof blues, but neither are conventional built-up roofs unless modifications are made or the entire system improved, including the deck and the supporting structure.

18. Better standards for materials are required, as well as performance criteria and test methods.

19. It is true that the quality of labor on a traditional roofing job is an important part of the whole because roofing is a labor-intensive trade. However, the level of technical competence required is not high and can be learned on the job under the direction of an experienced foreman. Due to sporadic terms of employment and less than desirable work, turnover is high and training must be repeated. This is likely to continue, even with single-ply roof membranes, but the errors made by the unskilled worker will result in much more trouble because of the drastically reduced margin of safety.

Reasons for Roofing Problems

Roof System Specifications

20. There are many ways to cut costs in BUR and most roofers know them. When faced with a difficult general contractor, bad weather, poor working conditions, or a tight competitive bid, a roofer can easily shave a job without the roofing inspector or the general contractor being aware of it. No one else on the job knows what the roofer is doing up on the roof and few care to look in case they demonstrate their lack of knowledge of roofing. There should be a reduced opportunity to skin the job with single-ply membranes, but this is offset by the increased need for applicator competence and understanding.

Development of Roofing Specifications

The earliest specifications for flat-roof membranes contained tar-saturated rag felt and pitch, or asphalt-saturated rag felt and asphalt. The second system often contained asphalt-coated felts with or without uncoated asphalt felts. A California manufacturer sometimes included an asphalt-saturated jute (burlap) fabric with several plies of rag felt. This made an exceedingly strong roof membrane. In the early 1930s in the Pacific Northwest asphalt roofs were built with a heavy 62-lb coated base sheet and three plies of 15-lb asphalt felt.

These old systems are described because after 35 years of trial and error, the 1930 roof with a base sheet and three plies of felt is more or less a standard system today. It grew out of attempts to eliminate the wrinkle cracking failures of roofs laid with saturated but uncoated felt over thermal insulation, and poured lightweight concrete and other types that contained a great deal of water.

The other method used to combat this type of failure is the taping of the joints in the insulation with glass-fiber tape 6 in. wide: 2 ft by 4 ft insulation requires 75 ft for each square of roofing; 3 ft by 4 ft insulation requires 58 ft. Tests on the effectiveness of taped insulation joints in preventing or reducing wrinkle cracking are inconclusive because of the multitude of felt, insulation, and bitumen combinations, and exposure to various interior and exterior conditions.

Another old standard roof, illustrated in Figure 10.1, contained one or two plies of nailed organic felt and three plies of mopped felt. Both asphalt- and tar-saturated felt were used on wood roof decks. Many of these proved unsatisfactory when moisture from green lumber and from the building interior caused the nailed felts to buckle, the tar-saturated felts being the most susceptible. The buckles were transmitted to the mopped felts, causing the flood coats of asphalt and pitch to slide off, taking the gravel with them and exposing the felt below. Tarred felt roofs deteriorated very rapidly owing to the cold flow properties of pitch. This type of failure increased as the rag content of the felt was reduced and the moisture-sensitive wood fiber felt was increased.

It is only recently that the two dry and three mopped type of roof system has been replaced by the asphalt base sheet and three asphalt mopped felts.

Priming

Plywood decks 2 and 3 should be primed with cut-back asphalt primer by the deck installer as soon as they are laid to keep moisture out of the wood. If possible, the edges should be primed while the sheets are still in the bundles. Use a roller coater or spray, 1 gal to 200 to 300 ft^2. Asphalt primer is suggested on decks 7, 8, 10, and 14 where felts or base sheets could be mopped directly to the roof deck.

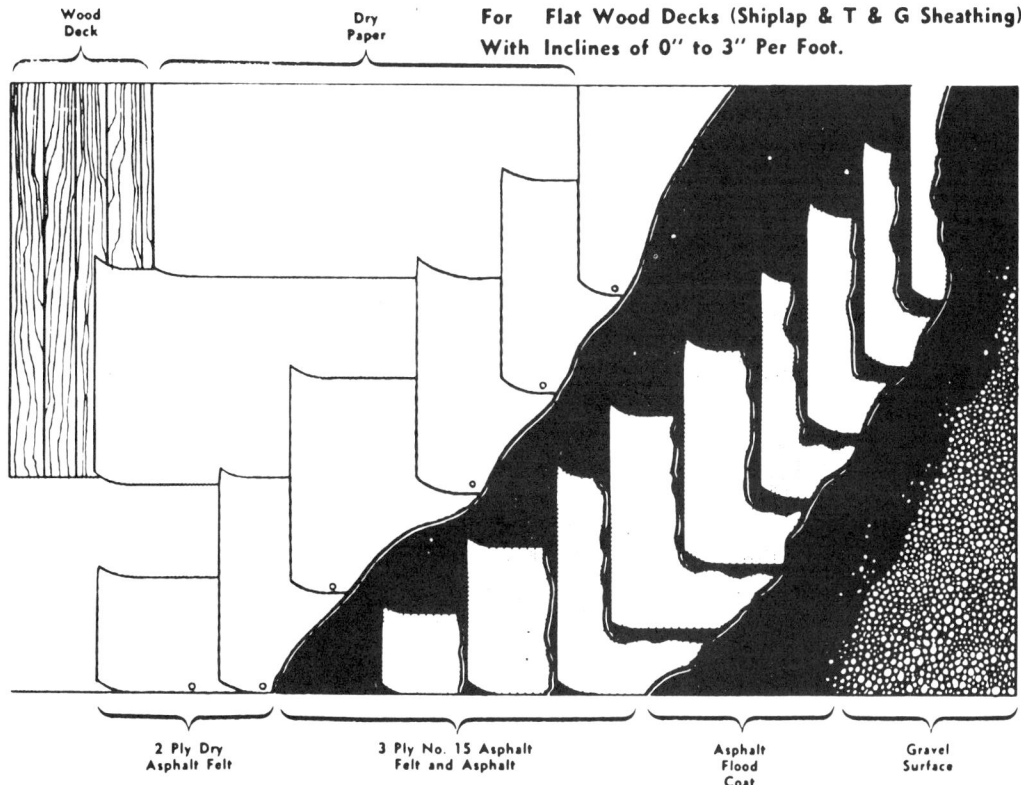

Figure 10.1   Old roof specifications for wood deck.

Coal-tar priming on decks 7 and 10 is suggested because of the possibility of dust on the surface preventing a good bond. The other decks or substrates suitable for tarred felts are 11 and 12. Neither contains portland cement.

Coal-tar and asphalt primers should be handled with care since they contain flammable and toxic solvents. Read the labels on the containers. Do not use in nonventilated areas. Priming coats must be applied to dry surfaces, but rain shortly after application will do them no harm.

Base Sheet (Asphalt Saturated)

Description
Organic wood fiber and rag, asbestos, and glass. Coated both sides with filled asphalt and separating or release agent, or antistick material. One or two squares per 36-in.-wide roll, 40 to 50 lb per roll.

Essential properties   Flexible fabric. Heavy asphalt coating with high ductility and minimum filler. High Mullen or bursting strength to resist nail pull-through. Good resistance to moisture absorption to avoid buckling. Inorganic felts are the most stable.

Roof System Specifications

Application Base sheets are generally laid one ply with 2- to 4-in. side laps, and nailed through nailing discs to nailable decks at approximately 12 in. on center in both directions. On wood and plywood decks a layer of unsaturated building paper is laid first and covered by the base sheet.

Where high wind speeds are expected, the deck should permit mopping of the base sheet with hot asphalt to prevent blow-offs. (See Table 4.1.)

Coated base sheets should be rolled out in warm weather and allowed to relax before nailing or mopping to the deck. Lightweight glass and combination sheet might be excepted.

If base sheets and ply sheets from different manufacturers are combined in the same roof system, their compatibility should be checked, as well as the compatibility with the mopping asphalt. Do not use organic base sheets or combination sheet on high-moisture-content decks such as 5, 7, and 9 through 12.

Base sheets should be covered as soon as possible with mopped felts, preferably the same day. They should not be used as temporary roofing. Cold-weather (below 40°F or 2°C) application is impractical owing to the inflexibility of the coated materials at low temperatures (see Figure 10.2).

Roofing Felt and Ply Sheets

Uncoated asphalt and tar-saturated organic felts (wood fiber and rag), uncoated asphalt and tar-saturated inorganic felts (asbestos), and asphalt-impregnated glass felts or mats are rolled directly into a mopped coat of hot bitumen applied to a moppable deck. (See columns J and K in Table 4.1.) Asphalt roofing felts are also mopped to base sheets and to insulation. Tarred felts are not appropriate over asphalt-coated base sheets; therefore, a pitch and gravel roof specification is only recommended to be fully mopped to deck 7 (poured concrete), deck 10 (lightweight concrete), or over insulation with a low capacity for bitumen absorption: minimum number of felt plies, four; gravel surface; maximum incline $1/2$ in. per foot (4.17 cm per meter). Four centimeters per meter is a good metric standard (i.e., 4%).

If a tarred felt roof is laid on decks 5 (poured gypsum), 9 (cellular concrete), 11 (vermiculite concrete), 12 (perlite concrete), or 13 (wood fiber and cement), two plies are laid with a 19-in. lap and nailed through caps with self-clinching nails, followed by three or four plies of mopped tarred felt laid in pitch and graveled over. The tarred felt does not offer the same resistance to moisture absorption from the deck or to wind suction as would an asphalt-coated base sheet.

For economy of labor, felts are laid parallel to base sheets and shingle mopped to each other. That is, they are lapped to achieve a two-, three-, or four-ply roof (see Table 10.1).

It is common practice to run felts across or at right angles to the slope of the roof so that water runs away from the laps and not toward them. This means that the roof is started at the lowest point as on a shingled roof. At the same time, since roofing felts are usually stronger in the machine direction than in the cross-machine direction, they should be laid across boards and at right angles to the long dimension of plywood, insulation layers, and any other sheet material. On flat or nearly flat graveled roofs, the second rule should apply, but on steeper smooth-surfaced roofs or mineral-surfaced selvage-edge types the first rule should apply because the laps are exposed.

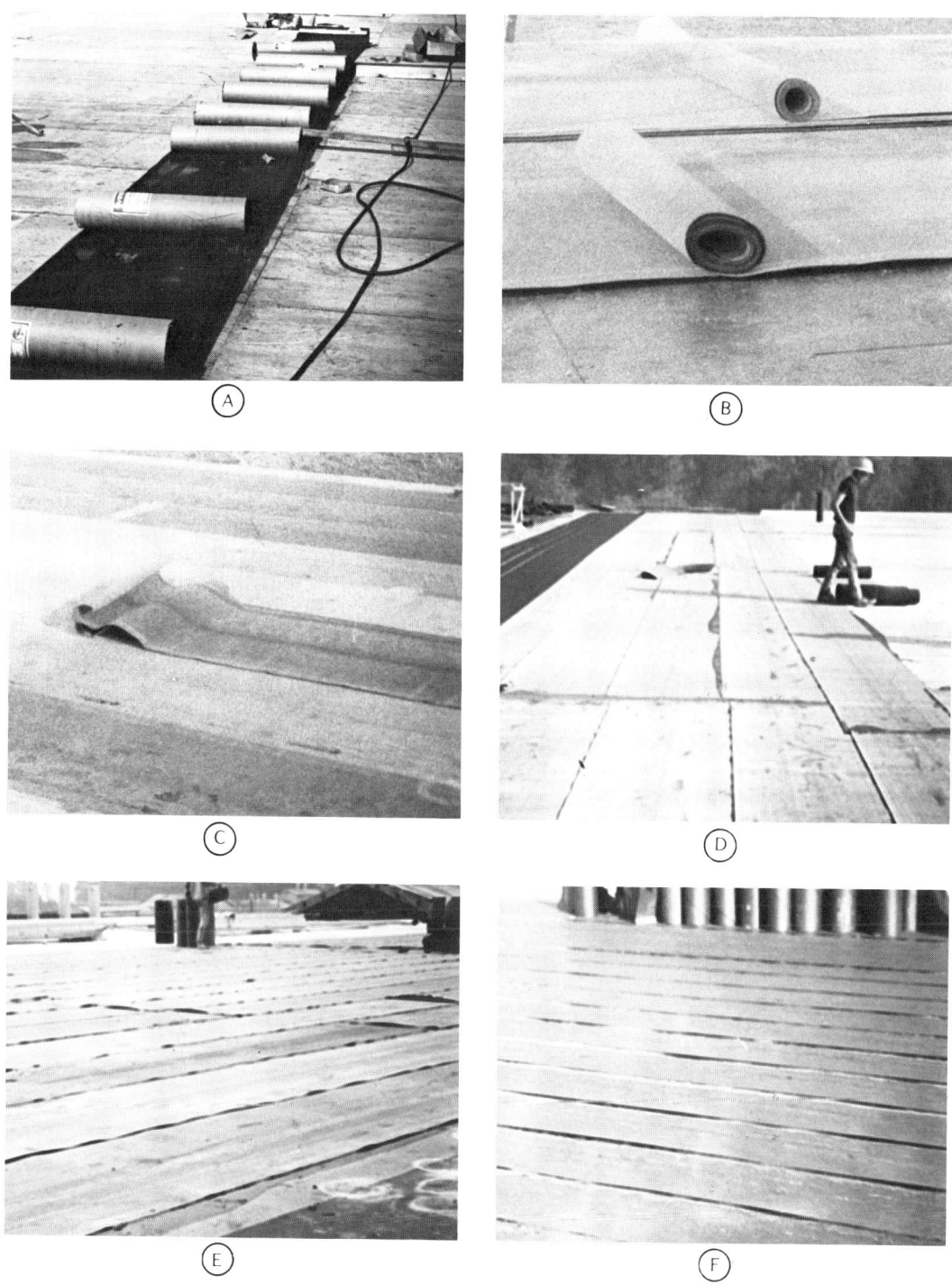

Figure 10.2 In spite of preexpansion (A), ends of rolls stay curled up (C) and edges are wavy (E). Even mopped felt layers don't lie flat (F). This material is organic felt laid in mild weather on a plywood deck.

Roof System Specifications

Table 10.1

	Felt 36-in. Wide		Felt 1-m Wide	
	Lap (in.)	Exposure (in.)	Lap (cm)	Exposure (cm)
Two ply	19	17	52.5	47.5
Three ply	24.66	11.33	68.34	31.66
Four ply	27.50	8.50	76.25	23.75
Head lap		2.0		5.0

Roofing membranes are subject to minute but constant changes in dimension due to temperature and moisture variations. The annual variation can easily exceed 200°F (93.33°C) and the diurnal variation 130°F (55.0°C). This means that a 100-ft width of organic felt and asphalt membrane, if free to move, will contract $7/16$ in. between 30° and 0°F and $1\,3/8$ in. between 0° and −30°F. This is a total of $1\,13/16$ in. (4.60 cm) for the 60°F increment.

Because of the differences in the coefficients of thermal expansion shown in Table 10.2, it is unwise to mix two types of roofing felt in the same roof membrane. For example, two plies of asphalt organic felt covered with two plies of asphalt asbestos felt can result in a separation of the two top plies from the two bottom plies, and buckles or wrinkles may form in the asbestos felts.

In the +30° to 0°F range the difference between an asphalt-coated organic, asbestos, or glass base sheet and tar-saturated organic felt ply sheet is too great to consider combining them in the same roof. Added to the problem of thermal expansion is the complicated problem of incompatibility of factory coating asphalts, saturating asphalts, and hot mopped coatings from different sources. A specially skilled laboratory technician is required to determine the compatibility of all the bituminous materials that are available.

There is great danger of slippage or separation of components in a roofing system when the components have different physical and chemical characteristics and when they are separated by a continuous unbroken layer of low-softening-point bitumen. Differences in temperature between the surface exposed to the sun and the underlying layers produce expansion and contraction stresses in the system, which

Table 10.2
Coefficients of Thermal Expansion for Roof Membranes

	Coefficient of Thermal Expansion per °F $\times 10^{-6}$			
	30 to 0°F		0 to −30°F	
Type of Membrane	L*	T*	L	T
Organic felt and coal-tar pitch	22.3	36.0	19.3	29.5
Organic felt and asphalt	2.7	12.6	13.9	37.4
Asbestos felt and asphalt	4.8	18.1	19.5	37.5
Glass felt (type 1) and asphalt	8.9	10.1	35.1	46.4

*L denotes longitudinal or machine direction of felts. T denotes transverse or cross-machine direction of felts. This can also be written MD for machine direction and CD for cross-machine direction.

can only be counteracted by secure, uniform attachments between felt plies, membrane to insulation or base sheet, insulation to vapor barrier, and vapor barrier to the deck. For further information on shrinkage of bituminous membranes, refer to R.G. Turenne, *Canadian Building Digest CBD-181*, July 1976, National Research Council — Division of Building Research, Ottawa, Ontario.

Back Nailing Felt

It is generally advisable to back nail horizontal asphalt ply sheets at approximately 12 in. on center 2 in. from the upper edge when the roof incline exceeds 1 in. per foot. Similar back nailing should be done when felts are run vertically or parallel with the slope of the roof. The purpose of the nails is to prevent the felt plies from sliding if the roof temperature approaches the softening point of the bitumen, which will increase gradually as the roof ages. All nails should be covered with not less than two plies of felt. It is not advisable to nail into or through low-density thermal insulation. Wood nailing strips in insulated roof systems have many disadvantages and are not recommended.

Roofing Bitumen: Asphalt and Coal-Tar Pitch

Asphalt Selection

The marketing identification of blown or oxidized asphalt is a simple type 1, 2, 3, or 4 in the United States, and type 1, 2, or 3 in Canada. The different types define the minimum and maximum softening point ranges and are shown in Tables 10.3 and 10.4. The selection by type number is generally related to the incline of the roof so that flow or sliding will be minimized at elevated temperatures. Recommendations by ASTM and CSA follow.

Table 10.3
Roof Inclines in Inches Per Foot

Asphalt	Type 1	Type 2	Type 3	Type 4
ASTM D312–1971	Max. 1.0	0.5 to 3.0	0.5 to 6.0	*
CSA A123-7 1973	Max. 0.75	0.75 to 1.5	Over 1.5	†

*For roofing with relatively steep slopes, generally in areas with relatively high year-round temperatures.

†Inclines are offered as a guide only and modifications of the inclines may be necessary for specific service conditions. CSA does not list a type 4 asphalt.

Table 10.4

Pitch Type	CSA A	ASTM A
Softening point	140 to 155°F (60 to 68.3°C)	129 to 144°F (54 to 62°C)
Incline	Max. 0.5 in. (Oct 1969)	3.0 in. with nailing 1.0 in. without nailing

Roof System Specifications

William C. Cullen of the National Bureau of Standards is reported to have stated, "Use the lowest melt point asphalt possible, commensurate with the slope of the roof." There is much to be said for this advice. Three reasons follow:

1. It takes less heating fuel and less time to raise the temperature to the equiviscous temperature (EVT).
2. The difference between the EVT application range temperature and the flash point of type 1 asphalt is greater than with types 2, 3, and 4, which reduces the possibility of fire.
3. Low-softening-point asphalts have better weathering properties than harder asphalts.

However, other factors must be considered in the selection of the appropriate asphalt softening point.

1. A substrate of high thermal resistance will increase the temperature of the roof membrane in summer.
2. A steep asphalt-surfaced roof without gravel or reflective coating can be heated above 200°F by solar radiation. The minimum softening point of type 4 asphalt is 205°F.
3. In southern latitudes or where the annual hours of sunshine exceed approximately 2,200 hours, an ordinary gravel- or slag-covered roof can have a surface temperature in excess of the minimum 135°F softening point of type 1 asphalt.
4. When applying a roof in hot weather, it may be impractical to use type 1 or 2 asphalt because workers and equipment stick to the roof before it is graveled. Some roofers start work earlier in the morning, but they run the risk of dew wetting the felt and deck surfaces. This moisture can be locked in the roofing system.
5. The final selection may be a matter of judgment on the part of the roofer as to which type of asphalt is used. Hard and fast rules on softening point related to roof incline cannot be made, and ASTM specifications reflect this conclusion.

Asphalt Quantities

The primary purpose of interply moppings is to adhere the felt plies together with a water-impervious layer of bitumen in a uniform thickness. It is imperative that there be no air spaces between the felt plies, particularly if they have been coated with asphalt at the factory.

Custom has dictated a mop coat weight of 20 to 25 lb per square regardless of the asphalt type, source, viscosity at the optimum mopping temperature, method of application, surface being mopped, or ambient temperature or wind-chill factor. These arbitrary figures, sometimes used unfairly to condemn an ordinary roof because it is underweight, are being challenged by an awareness that all asphalts do not have the same characteristics, since they originate in several states and several countries. There are also mixtures of more than one kind of asphalt flux.

New standards may be in force before this text is published. The standards will specify the equiviscous application temperature range for all grades, types, and kinds of asphalt available for roofing. The equiviscous temperature (EVT) will relate to the viscosity of the asphalt at various mopping temperatures. Heating the asphalt before application will relate to the EVT and the flash point of the material, and also to the finished blowing temperature. These refinements will provide more realistic controls that will result in proper adhesion and proper bitumen thickness, at the same time reducing the chance of damage to the asphalt and the possibility of kettle fires.

Pitch Selection

According to ASTM and CSA there is only type A roofing pitch available, and it is suggested by CSA that the maximum roof incline be 0.5 in. per foot. On the other hand, ASTM D450-71 suggests 1.0 in. per foot without nailing and 3.0 in. per foot with nailing. With all due respect to ASTM, the author believes that the CSA recommendations are more realistic. In October 1969, CSA reduced the incline for pitch from 1.5 in. to 0.5 in. per foot.

The National Roofing Contractors Association (NRCA) shows a maximum of 0.75 in. with back nailing above 0.5 in. The Canadian Roofing Contractors Association shows the maximum incline for pitch and gravel roofs at 0.5 in. per foot.

Pitch Quantities

The correct theoretical weight of pitch between felt plies and the flood coat is 20% more than asphalt because of the greater specific gravity of pitch or lesser volume for the same weight. Interply moppings are 25 to 30 lb and flood coats 75 to 80 lb per square. The temperature of the pitch will affect the coating thickness to a considerable degree, as will the porosity of the surface on which it is applied.

Surfacing Materials

The purpose of the top surfacing on a hot-applied built-up roof is to protect the underlying felts from moisture, solar radiation, wind erosion, and roof traffic.

Tarred felt roofs are always finished with a 75 lb poured flood coat of pitch and approximately 400 lb (33.75 kg) per square (100 ft^2) of clean opaque gravel or 300 lb of slag. The gravel may be round or crushed, and both gravel or slag must be opaque to ultraviolet light to avoid degradation of the flood coat and loss of the gravel. White or colored stone is not suitable for pitch roofs because of the staining.

Asphalt felts (organic, asbestos, and glass) are generally graveled as above, using a 60 lb (29 kg) flood coat of asphalt per square. The recommended maximum incline for a graveled roof is 3 in. per foot, but under certain conditions where a layer of opaque white chips can be held by type 4 asphalt or by a primary surfacing of crushed rock, the incline can be raised to 4 in. Limestone chips and other nonopaque rocks should not be embedded directly in a flood coat of hot asphalt. Oxidation of the asphalt will destroy the bond between the two. Inorganic asphalt felts on roofs above 0.5 in. per foot can be surfaced with a mop coat of asphalt, type 2, 3, or 4, but they should be covered with a heat-reflecting coating in all cases. The coating must be maintained in a clean bright condition at all times. It is not suitable for areas with unusual air pollution conditions that reduce reflectivity.

Roof System Specifications

Figure 10.3 This is not a common gravel or stone surface for a built-up roof, but the texture and color suit the building. Such a surface might be useful in discouraging unauthorized traffic on school roofs, but the stones would make dangerous missiles. Ceramic-coated gravel is available in several large sizes in California.

An asphalt emulsion coating can be used, but only on a factory-coated inorganic felt roofing or felts that have been hot mopped with asphalt at the site.

Cut-back asphalt coatings are not recommended for new roofs.

Special Note 1: Types 3 and 4 asphalts combined with organic felts, smooth surface, are not recommended because of the possibility of excessive blistering. Uncoated glass felts, on the other hand, may require types 3 and 4 asphalts to avoid deep penetration of the asphalt into the porous felt. Glass felts made in certain ways have been known to float up on soft asphalts, leaving the asphalt on the bottom and the glass on top. A factory-coated glass felt may help to avoid this.

Special Note 2: A double flood coat and extra graveling are sometimes specified for extreme conditions of exposure. This is practical only if all loose gravel in the first layer is removed before the second flood coat is applied. This is not easy to do with round gravel and impossible to do with crushed rock.

Special Note 3: A mineral-surfaced cap sheet of organic or inorganic felt laid separate to the mopped plies of felt is not rcommended owing to the possibility of blistering.

Types of Roof Decks

1. Boards on wood joists or purlins. Square edge, shiplapped, or tongue and groove (T&G).
2. Plywood on wood joists or purlins.

3. Plywood over T&G decking.
4. Steel, with plywood, gypsum board, or insulating board overlay.
5. Poured gypsum.
6. Precast gypsum.
7. Poured concrete.
8. Precast concrete with insulation or concrete fill over.
9. Cellular concrete.
10. Lightweight concrete.
11. Vermiculite concrete.
12. Perlite concrete.
13. Wood fiber and cement slabs.
14. Asbestos cement cavity decks, with insulation or concrete fill over.
15. Thermal insulation on structural deck.

Typical Roof Membranes

Materials	Deck Nos.	Incline
Organic base sheet, nailed. Three plies No. 15 asphalt felt. Asphalt flood coat and gravel. Approx. total asphalt weight per square, 120 lb (54 kg)	1, 2, 3, 6, 13	0 to 3 in. per ft: notes A, B, C, D, F
Organic base sheet mopped to deck. Three plies No. 15 asphalt felt. Asphalt flood coat and gravel. Approx. total asphalt weight per square, 140 lb (63 kg)	2, 3, 4, 7, 8, 14, 15	0 to 3 in. per ft: note D
Four plies No. 15 asphalt organic felt. Asphalt flood coat and gravel. Approx. total asphalt weight per square, 140 lb (63 kg)	2, 3, 4, 7, 8, 14, 15	0 to 3 in. per ft: note D
Four plies No. 15 tarred organic felt. Pitch flood coat and gravel. Approx. total pitch weight per square, 175 lb (79 kg)	4, 7, 8, 14, 15	0 to 1/2 in. per ft
Asbestos base sheet, nailed. Three plies No. 15 asphalt felt. Asphalt flood coat and gravel. Approx. total asphalt weight per square, 120 lb (54 kg)	1, 2, 3, 5, 6, 9, 11, 12, 13	0 to 3 in. per ft: notes A, B, C, D, F, G
Asbestos base sheet mopped to deck. Three plies No. 15 asphalt asbestos felt. Asphalt flood coat and gravel. Approx. total asphalt weight per square, 140 lb (63 kg)	2, 3, 4, 7, 8, 10, 14, 15	0 to 3 in. per ft: note D, G
Four plies No. 15 asphalt asbestos felt mopped to deck. Asphalt flood coat and gravel. Approx. total weight of asphalt per square, 140 lb (63 kg)	2, 3, 4, 7, 8, 10, 14, 15	0 to 3 in. per ft: note D, G

Roof System Specifications

Typical Roof Membranes (cont'd)

Materials	Deck Nos.	Incline
Asbestos base sheet, nailed. Three plies No. 15 asphalt asbestos felt. Mopped coat of asphalt. Reflective surfacing. Approx. total asphalt weight per square, 90 lb (40.5 kg)	1, 2, 3, 9, 13	3 to 6 in. per ft: notes A, B, C, D, F, G
Four plies No. 15 tarred asbestos felt. Pitch flood coat and gravel. Approx. total pitch weight per square, 175 lb (79 kg)	4, 7, 8, 14, 15	0 to 1/2 in. per ft: note G
One ply No. 15 asphalt felt. Two plies mineral-surfaced selvage-edge roofing, lapped 19 in. and nailed. Approx. weight type 3 or 4 asphalt per square, 40 lb (18 kg)	1, 2, 3, 6, 13	1 to 6 in. per ft: notes E, F
Coated glass base sheet, nailed. Three or four plies glass ply sheet. Asphalt flood coat and gravel. Approx. total weight asphalt per square, 150 to 180 lb (67.5 to 81 kg)	1, 2, 3, 5, 6, 9, 11, 12, 13	0 to 3 in. per ft: notes A, B, D, F
Coated glass base sheet, mopped to deck. Three plies glass ply sheet. Asphalt flood coat and gravel. Approx. total weight asphalt per square, 180 lb (81 kg)	2, 3, 4, 7, 8, 10, 14, 15	0 to 3 in. per ft: note D

Notes

A: Add one layer of unsaturated building paper stapled to deck 1.

B: Nail base sheet through flat nailing discs at 12 in. on center in both directions.

C: Recommended for warm-weather application, i.e., above 50°F (10°C) so that the coated base sheet will be flexible and will lie flat.

D: Back nail mopped felts above 1 in. per foot incline on nailable decks. Use self-locking nails in low-density materials.

E: Inorganic felts are preferred with white or light-colored granules on the mineral-surfaced portion to reduce the surface temperatures.

F: For nailable decks only.

G: Refer to chapter 6 for availability of asbestos roofing materials after 1983.

Protected Membrane Roofs

In this system the primary waterproofing membrane is placed on the roof deck and is covered with thermal insulation and a heavy protective ballast. The separate vapor barrier is eliminated and the insulation is not sandwiched between vaportight membranes. Since the membrane is below the insulation, it is protected from solar radiation and mechanical damage and temperature extremes. In a roof system where the membrane is above the insulation, the temperature of the membrane differs by 180°F between −20° and +160°F (−29° and +71°C), but in the protected system the temperatures are narrowed to 30°F between +55° and 85°F (13° and 29°C). These figures assume a thermal resistance of approximately 8.0 above the membrane and 1.18 for a concrete roof deck and inside air film below the membrane. The waterproofing membrane and the roof deck are more or less in the same environment.

Protected Membrane Roofs

In the existing "insulated roof system," the deck is the only part that is insulated. The roof membrane remains in a hostile and damaging environment.

The components in the protected membrane system are

1. *Roof deck:* This can be any of the 14 decks in "Types of Roof Decks" provided they are strong enough to carry an extra 15 to 20 lb per square foot dead load imposed by the roofing, insulation, and ballast. An ordinary graveled dead-level roof has 4 lb of gravel and may occasionally carry several inches of water, ice, and snow. As long as the deck is sloped to drain and the regular gravel cover is omitted from the roof membrane, the extra strength required is minimal. Steel decks must be covered with plywood or gypsum board to support the roof membrane. A positive slope to drains is recommended to reduce the flotation effect of ponded water on the insulation and to reduce the absorption of water by the insulation. With decks having relatively high thermal resistance, such as decks 5, 6, 9, 11, and 12, the resistance in the deck should not be more than one third of the total in the system. Expressed another way, the resistance of the insulation above the roof membrane should be at least double that in the roof deck. If this rule is reversed, there is a possibility of condensation at the roof deck-roof membrane interface.

2. *Membrane:* Although the protected membrane system is historically very old it is still more or less in the development stage as far as modern buildings and materials are concerned. Three types of membranes are being used.

    a. A built-up roof using organic and inorganic felts and hot asphalt.

    b. Thin (49 to 60 mils) sheet materials such as polyethylene, polyisobutylene, butyl rubber, neoprene, and PVC.

    c. Fluid-applied epoxide vinyl polymers, neoprene, and rubber-asphalt compounds.

    Since the waterproofing membrane or coating is in intimate contact with the deck, it must be able to accommodate any movement in the deck without rupturing. Therefore, changes in dimension of the deck elements must be predictable considering the environment in which it is located. Even if the membrane is not firmly attached, the weight of the ballast increases the friction attachment. Precautions must be taken to prevent some fluid-applied systems from being forced up through the joints in the insulation or down through cracks in the deck. The membrane must be able to withstand being in a wet environment, that is, below a relatively vapor-impermeable insulation for long periods in some climates. It must also not be damaged by vegetable growth and airborne chemicals.

3. *Insulation:* The principal material being used is an extruded polystyrene; however, others may prove suitable in certain environments. It is important that the insulation be laid in single thicknesses and not in multiple layers. The insulation will be exposed to moisture and freeze-thaw conditions in some areas. Wetting and drying must not damage it. A high degree of thermal resistance is required to allow for some reduction due to moisture absorption and to prevent

Roof System Specifications

the need for excessive thicknesses, which result in a large crack volume and certain difficulties at flashing junctures and drains. If polystyrene is used, the flotation force upward is approximately 5 lb per inch of thickness when completely immersed in water. This force must be counteracted by increased ballast weight. The bond between the insulation and the roof membrane must therefore be considered. Any waterproof coating on the insulation should not prevent evaporation of moisture from the edges and top surface. A complete envelope would not be advisable.

4. *Ballast:* Washed, round, opaque gravel free of fines and vegetable matter 3/4 to 1 1/4 in. in diameter and 1 1/2 to 2 in. deep provides reasonably good cover for the insulation and a dead load of approximately 11.5 to 15.4 lb per square foot (based on 92.6 lb per cubic foot). In addition to the ballast required, the gravel or any other material must protect the polystyrene from solar radiation and mechanical agents, and allow the system to lose moisture by evaporation. Crushed rock would reduce the rate of evaporation, but would resist movement by foot traffic. Round gravel is preferred. Concrete paving slabs laid loose have also been used as ballast and radiation protection. Dense concrete 1 1/2 in. thick weighs 17.5 lb per square foot. These slabs make an excellent surface that cannot be moved about by foot traffic or wind, but can be removed for inspection of the insulation. Slabs are recommended only when the roof slopes to drain. Large areas of poured-in-place concrete are not recommended. Small concrete paving slabs with cast-in-place legs or pads 1/2 to 1 in. high provide an excellent drainage space on top of the insulation, and are a great asset in venting the system. This would prevent the rise in temperature of the slabs due to the heat-sink effect from deteriorating the insulation. Polystyrene begins to deform at about 150°F and to reduce in volume at 170°F. A light-colored surface on the slabs would be useful in controlling heat absorption, but it is not likely that the surface could be kept clean.

Appropriate Applications

The principles behind the design of the protected membrane system eliminate many of the problems with the vapor barrier-insulation-roofing sandwich, but the selection and design of the system requires careful consideration to satisfy the conditions present on individual buildings (see Figure 10.4).

Figures 10.4 and 10.5 show the construction of a protected membrane roof on a 28-story office building. The structural concrete deck was covered with a lightweight concrete fill for drainage. The deck was primed and covered with a 55-lb coated asbestos base sheet and four plies of No. 15 asbestos felt laid two and two and mopped with asphalt. One inch of extruded polystyrene insulation was laid in cooled asphalt, and 1 1/2-in. concrete paving blocks covered the insulation. Figure 10.6 illustrates the important differences between a standard or conventional roof system and the protected membrane system.

Figure 10.7 is a graph representation of the approximate temperatures in a roof system under steady state conditions. The horizontal lines are drawn to any scale and represent the thermal resistance of the various components. The vertical lines are also drawn to any scale and represent the outside air temperatures at the top and the inside

Figure 10.4 Construction of a protected membrane roof. Asbestos felts are mopped in by hand because of the 44-ft. building width and because construction is barely ahead of the roofing. The pairs of concrete pedestals support window cleaning rails (E). Roof ready for insulation (F). Note the portion of the roof still not ready for roofing. Insulation and paving blocks being installed, with other blocks stored on temporary supports

Figure 10.4 (cont'd) (G). Part of the roof deck at right (H) are not yet filled in. Openings were left for heavy equipment to be installed on the floor below. After the roof was completed, the action of steel erectors continued. One crane of two on the roof is at the left and steel rails at the right (J). The roof membrane is protected below the concrete paving blocks and the insulation (K).

Figure 10.5 Completed roof. (A–C) Raised sections are removable for machinery changes. Flashings are stainless steel. (D) Completed building. Small white squares are drain outlets. Large white rectangles are removable sections roofed on asbestos cement cavity decking. Center penthouse houses elevator machinery, and the roof serves as a helicopter landing pad in emergencies. (Photo courtesy of MacMillan Bloedel Co. Ltd.)

air temperatures at the bottom. The numbers for the thermal resistances can be extended upward and the outside temperatures reduced well below 0°F as they are indicated in this drawing. Temperatures within the system can be roughly determined for both winter and summer conditions. They are not exact and only show a condition in an instant of time. In reality the temperatures in a roof system are constantly changing, sometimes slowly and sometimes quickly, depending on the exterior environment and the reaction of the components in the system. This explains in part how flexible roof coverings can be distorted to failure so slowly that the reason or reasons are often difficult to ascertain.

In Figure 10.7, the winter temperature curve is represented by the line at the left starting at 70°, the interior air temperature, to the intersection of vertical line 0 which is 0°F, and the horizontal line 9 which represents the total resistance value of the interior air film (0.61), a steel deck and vapor barrier (0.39), insulation (8.0),

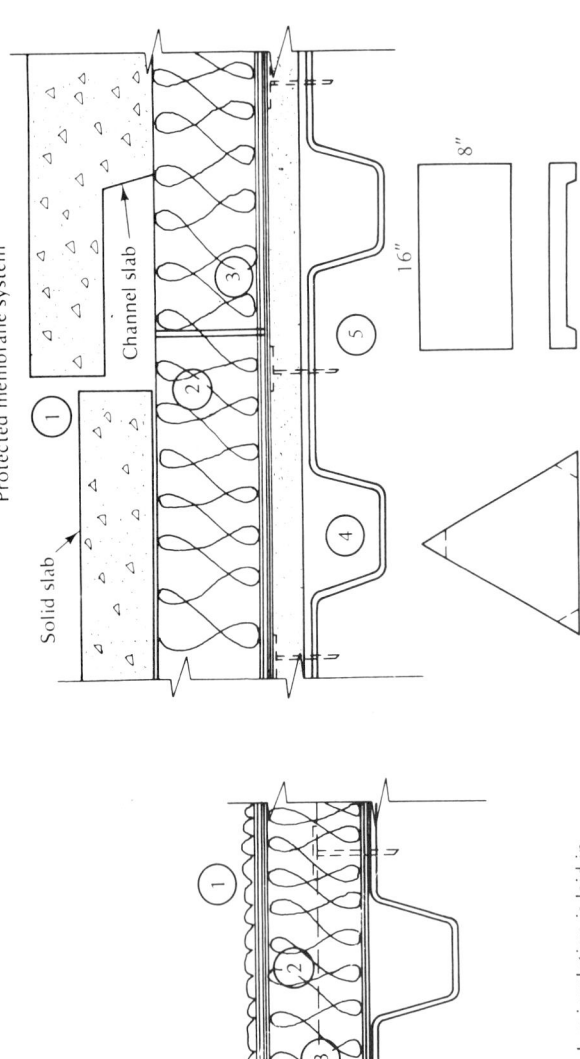

Standard system
(insulation sandwich)

Steel deck

Ⓐ

Protected membrane system

Solid slab

Channel slab

Ⓑ

1. Roof membrane exposed to weather and traffic.
2. Air-filled insulation and joints provide storage for trapped moisture entering from above or below. Compression of soft insulation can puncture a felt roof at the nails. Expansion and contraction of insulation destroys roof. Density of insulation must permit nailing on steel decks. Few comply. Deck and roof are in different environments.
3. Unless insulation is laid in two layers with nails or screws in the first layer, the insulation must be moppable with hot asphalt. Plastic foams are impractical.
4. Heavy insulation layers require long fasteners.
5. Vapor barrier and insulation are not well supported over open flutes. Fire resistance is minimal.

1. Traffic surface and protection for entire roof system. Ballast weight improves dead load/live load ratio. Channel type slabs improve drainage and breathing, and reduce solar heat gain by insulation.
2. Insulation protects roof membrane, substrate, and deck. Moisture vapor not trapped in airtight sandwich.
3. Combination roof membrane and vapor barrier on solid base. No gravel.
4. Gypsum board provides fire proofing and support for roof membrane.
5. All fasteners are short, and none passes through the insulation. Wind damage on steel decks is virtually eliminated

Figure 10.6 Comparison of standard roof system with protected membrane system.

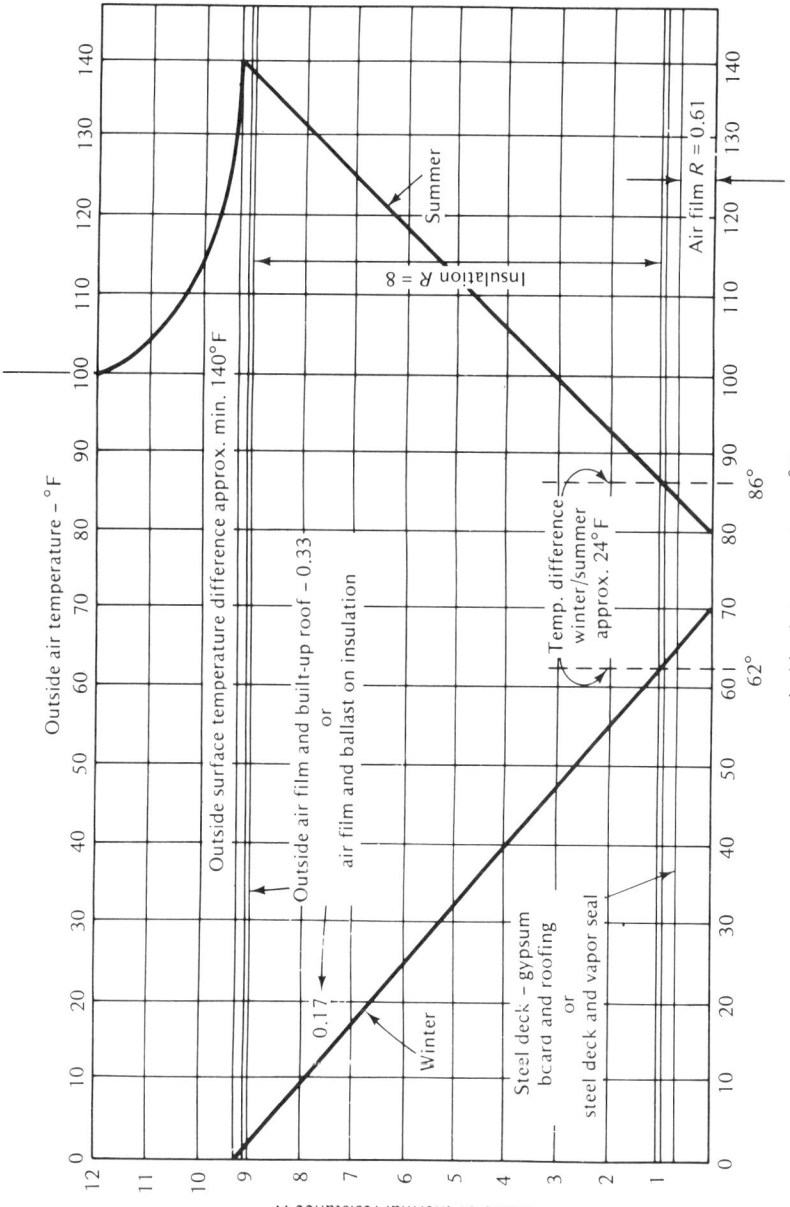

Figure 10.7 Temperature curves for roofing systems to serve as a guide to temperatures within the system under steady-state conditions.

Roof System Specifications

and the roof covering and exterior air film (0.50). The total thermal resistance is 9.0 to 9.5. Where the diagonal line intersects the horizontal lines, the temperature can be approximated by dropping a line to the lowest line. In this case the temperature at the vapor barrier is 62°F. If more insulation is added, the temperature line assumes a steeper slope and the temperature at the vapor barrier is increased. Also if the line is extended to $-30°$, the vapor barrier temperature will drop until it could be below the dew point of the inside air.

The summer curve at right is not quite as useful but does demonstrate that if the roof membrane is located below the insulation as in the Protected Membrane System, the difference between winter and summer temperatures is only about 24°F. However, if the roof membrane is on top of the insulation, the difference can be 140°F or more. This wide temperature range, changing constantly daily and seasonally, is believed to contribute to the destruction of the membrane. It also seriously affects the insulation, especially if moisture is present in the system and cannot escape.

This method of approximating temperatures is useful when the deck itself has a high thermal resistance or when the thermal characteristics of the system are to be changed by adding insulation either above or below the deck.

It will be seen that in a conventional roof system, if the air/vapor barrier is omitted, the moisture vapor might migrate, due to a vapor pressure differential, to the roof membrane itself or to a level in the insulation layer that is below the dew point. This is why absorbent insulation becomes wetted at the top first, and as the resistance value is reduced, the wetting travels downward.

11

Steel and Aluminum Roofing

Advantages

The principal advantage of metal roofing is its ability to repel water. It is lightweight, rigid, available in many shapes and color, is not degraded by solar energy, is easily worked, and can be installed quickly in any weather, especially in wet weather when built-up roofing materials cannot be installed. It is adaptable to many building designs and has a relatively long life.

Many of the problems associated with thermoplastic flexible roof systems used together with thermal insulation are eliminated with metal roofs. The installation by sheet metal and steel erectors is more precise and is a more skillful trade than built-up roofing and the results are generally more predictable. One decided advantage is the positive incline that must be present to drain off water.

The metal roofing described in this chapter is exposed to the weather and is the primary rain-shedding device. It must be suitably protected from corrosive elements by galvanizing, terne coating and painting, factory enameling, or mastic coatings. Aluminum and various copper alloys or lead-coated copper are also used. Proper allowances for the expansion and contraction of these metals are made in the fastening design and execution, both for end and side laps. Very long lengths without jointing can be obtained from some manufacturers who have developed special interlocking

Steel and Aluminum Roofing

and machine closing standing ribs. Exposed fastening of short lengths should be avoided because of the inevitable loosening of the fasteners, leaking, and expensive maintenance.

Roof Decks

Corrugated, shaped, or deformed steel and aluminum roofing is usually laid over purlins or spaced sheathing since the contours of the metal are able to span open spaces and to carry loads. It is important to differentiate between metal sheathing for side walls and for roofing. The correct spacing of supports is generally specified by the manufacturer for specific roofing shapes and metal gauges, but this should be adjusted for any unusual live load that may be expected due to local conditions. Metal can also be laid on solid sheathing; the advantages of this are discussed later.

Insulation

When insulation is used in a metal roof system it is located below the metal, which leaves the metal exposed to the weather. The designer must consider the building requirements and the interior and exterior environments before selecting the type, thickness, and location of the insulation and air/vapor barriers. There is the possibility of thermal bridges from the interior to the metal roofing through the fastenings and girts, which could be troublesome because of spot condensation in extreme conditions. There is also the possibility of condensation on the underside of the metal roofing, particularly on aluminum or copper because of their rapid conduction of heat. Aluminum conducts heat 4.5 times as fast as steel and 1,830 times as fast as soft wood. Copper conducts heat nearly twice as fast as aluminum. This is why copper and aluminum are harder to solder than steel. The nighttime temperature of metal roofing will depend to some extent on color and weather conditions. A dark color will be 10° to 15°F below ambient temperatures when the sky is clear and the roof has no snow cover.

Vapor Barriers

In cold climates, great care must be taken to provide air/vapor barriers below the insulation (all kinds) to keep water vapor away from cold metal. A ventilation system between the insulation and the metal would help dissipate vapor escaping from the interior and prevent moisture from the outside air entering the system and condensing on the underside of the metal during low nighttime temperatures. In freezing temperatures it is relatively easy for ice to form under the metal and to drip back into the building when the sun heats up the roof surface and melts the ice.

Under certain conditions, it may be advisable to locate a layer of plywood of waferboard above the insulation to support a layer of resin-sized paper which acts as an air barrier for the metal roofing. The vapor barrier is still required under the insulation. This may appear to be a complicated system but if it can remove the problems that persist with built-up roofing systems it will be a step ahead.

An existing built-up roof suitably prepared and with positive slope can form the basis of a new metal roof, plus additional thermal insulation.

Figures 11.1, 11.2, 11.3, 11.4, and 11.5 show details of metal roofs.

Figure 11.1 Example of steel roofed building in a cold climate.

121

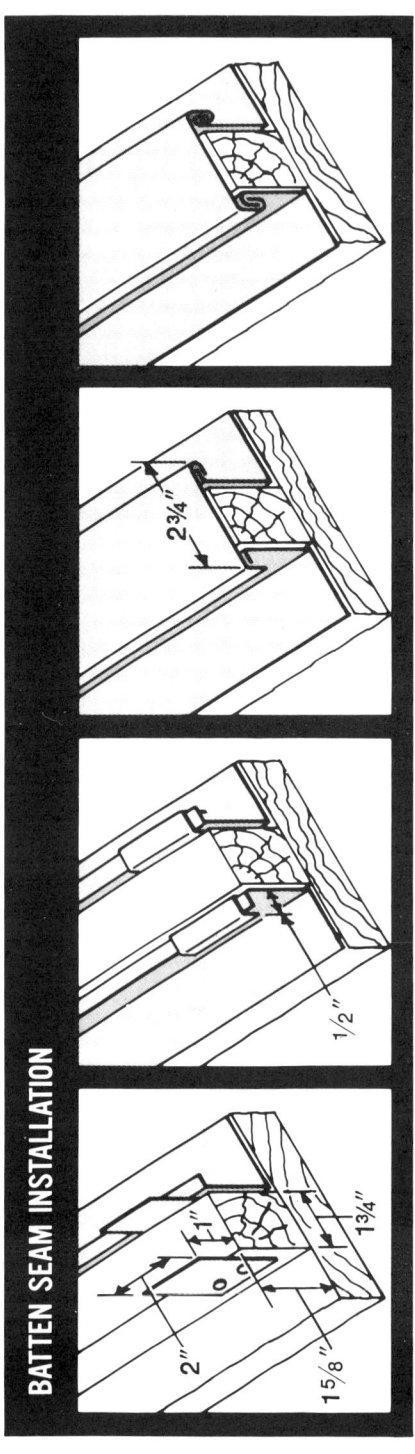
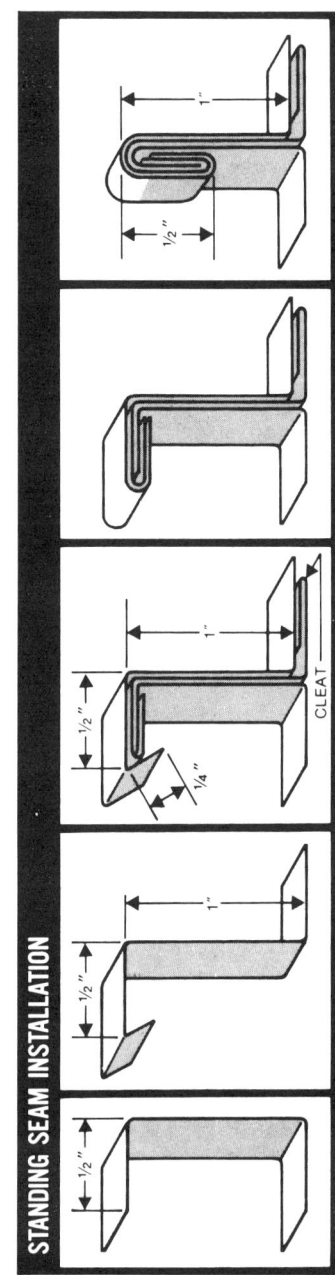

Figure 11.2 Details of batten seam joint using hand method of installation, and standing seam joint either hand or machine formed. Follansbee Steel Corporation.

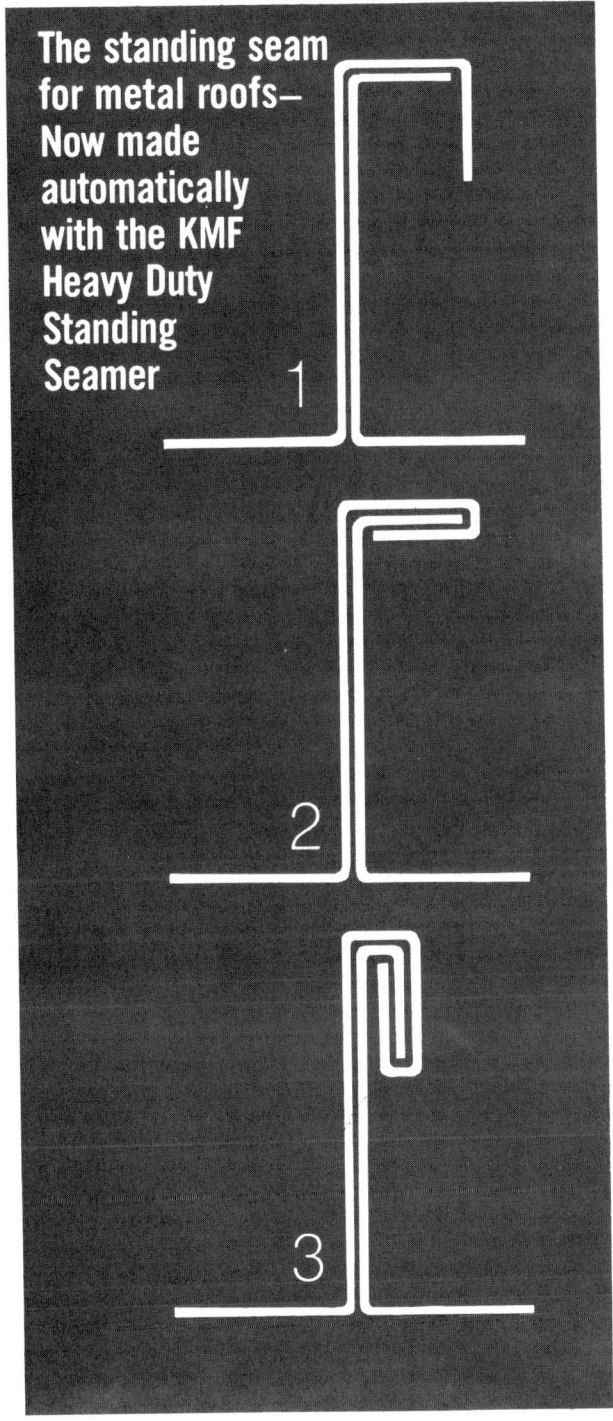

Figure 11.5 Details of seam formation. KMF Equipment.

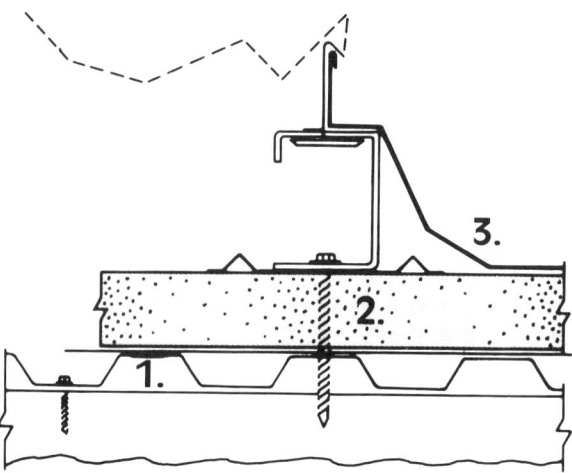

Figure 11.3 Standing seam joint. Bethlehem Steel Corporation.

The 10 ft. sheet, prefabricated in ELT's shop, is laid onto deck in preparation for seamer.

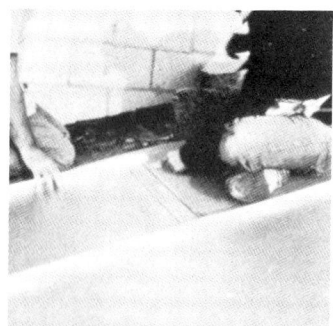

The clips, also prefabricated on a jig, are set into place.

A mechanic carefully aligns panel.

This small machine is responsible for big savings. The standing seamer produces standing seams at a constant rate of 12 ft. per minute.

Machine Manufacturer
KMF Equipment Corp.
Philadelphia, Pa.

Figure 11.4 Illustration of mechanically formed standing steel seam.

12

Drainage Systems

A dead-level roof is invariably drained by inside cast-iron or sheet-metal (aluminum, stainless steel, copper, or galvanized iron) drains that are connected to an underground sanitary or storm sewer system. This type of roof never drains completely and much of the water disappears by evaporation. The evaporation of water can leave concentrated solutions of destructive chemicals on the roof, which may seriously affect the roof membrane. All projections through the roof as well as the perimeter flashings must be carefully flashed, because the water depth cannot be accurately predicted. All penetrations of the membrane are potential leaks.

There are several manufacturers of excellent cast-iron drains, each making a variety of styles and sizes. Josam and Zurn are two, both cast iron, and both reliable.

Some miscellaneous notes on drains and drain installation follow for level roof areas.

1. The number and size of inside drains should be determined by the building designer. The method used will vary geographically and be influenced by local building regulations and other factors relating to building design. One method uses as the basis of design the 15-minute maximum rainfall that will be exceeded on the average once in 10 years. The diameter of the leader pipes in inches is related to the hydraulic load from the roof. This is defined as the maximum 15-minute rainfall in inches multiplied by the sum of the roof in square feet and one-half the area in square feet of the largest adjacent vertical surface. (Refer to environmental data services.)

Drainage Systems

2. Other useful information on the rate of fall related to gallons per minute and per hour is contained in drain manufacturer's catalogues.
3. All drains should be installed with a lead flashing flange secured to the drain sump and located on top of the roofing felts. Flanges should be set in asphalt gum, nailed and at the outside edges, and covered with two extra plies of felt and asphalt before the roof is graveled. (See Figure 12.1 for a typical drain installation on a concrete deck.) The flange should extend not less than 6 in. beyond the outside of the strainer ring. The flange may not be supplied with the drain.
4. Unless a lateral pipe is run from the drain to the vertical stack, an expansion sleeve should be placed between the drain outlet and the vertical stack. Lateral runs should have accessible cleanouts.
5. Wherever possible, locate drains at the center of roof deck spans.
6. Do not locate drains near columns, bearing partitions, exterior walls, penthouses, skylights, roof hatches, expansion or control joints, or any other projection through the roof.
7. On any individual roof area, use no less than two drains or at least one drain and one emergency overflow scupper. Scuppers should be placed in all parapet walls below the upper edge of the base flashings.
8. On schools, equip all inside drains with theft-proof and shatter-proof drain strainers. Roofs sloped to outside gutters are generally more satisfactory.
9. The use of two or more smaller drains, but not less than 3 in. in diameter, is safer than using one large drain: 3 in. diameter $= 4.71$ in.^{2}; 4 in. diameter $= 12.56$ in.^{2}; 6 in. diameter $= 28.26$ in.^{2}.
10. In areas where the volume of airborne waste is high, such as from wood-processing plants and flour mills, strainers can be eliminated and drains run to grade rather than to internal piping.
11. The use of controlled flow drains that are equipped with specially designed fixed or movable restrictive devices called weirs is not looked on with favor by some roofing associations, because they feel that savings are being made in the piping system at the expense of the roof. It is felt that a potential water depth of 6 in. is too dangerous unless unusual steps are taken in the design and construction of the roof deck, membrane, and flashings, plus extra-large overflow scuppers at parapets 2 in. above the roof level. It is worth noting that a column of water has a hydrostatic pressure of 0.4344 lb per square inch for each foot of head. Therefore, the roof water on a building 40 ft high has a potential hydrostatic pressure of 17.376 lb per square inch or 2502.144 lb per square foot.
12. Slightly sloped roofs can be drained with internal drains located at the low points, or to scuppers in parapet walls, or low curb cant strips, or to outside gutters.
13. Cast-iron drains are preferred to light sheet metal, except stainless steel for special conditions. The base of the strainer should be at deck level. (See Figure 12.1 for installation on concrete deck.)

Figure 12.1 Installation of drain sump and lead flashing flange on protected membrane roof system.

13

Metal Flashings

A flat or low-sloped built-up roof can be described as a flat tray with slightly raised edges that catches and diverts rain water to the drainage outlets. Where the roof terminates at open roof edges or at walls, the connection is made with a base and counter flashing, usually not physically connected, in order to allow for movement of the deck or wall, or both. Base flashings may be constructed of metal or roofing fabric. Counter flashings are generally metal. Where projections occur that penetrate the roof covering, the element may have a base flashing of metal secured to the roof and a metal counter flashing secured to or covering the projection. Large openings, such as skylights or ventilator ducts, are better flashed to a curb several inches above the roof level.

Basic Rules for Flashings

1. Use flashings to divert water to the roof and to drainage outlets.
2. Design flashings so they do not hold water. Slope wall copings back toward the roof area.

Metal Flashings

3. Keep ferrous metal flashings out of ponded water.
4. All exposed and concealed metal edges should be folded back 1/2 in. in a safe or seamed edge.
5. If possible, avoid flashing widths or girth more than 24 in. (60.96 cm) in one piece.
6. When using galvanized iron, specify a minimum zinc coating of 1.5 oz per square foot, and paint both sides with two coats of asphalt paint plus a light-colored reflective coating on all exposed surfaces after installation. This reduces surface temperature and some movement due to heating and cooling.
7. As an alternative to item 6, use a factory-enameled galvanized sheet where soldering is not required.
8. Avoid connecting nonferrous metals in long lengths to the roof membrane (e.g., copper, aluminum, and zinc or zinc alloys), either above or below the roof membrane.
9. Avoid using dissimilar metals in close proximity or where they can be connected by water flow. The farther they are apart on the electromotive force series, the more chance there is for electrolytic action in the presence of an electrolyte such as water or certain chemicals. The greater the distance between the electromotive potentials, the greater the chance for corrosion of the metal with the higher potential.
10. Use annular ring nails or screw thread fasteners concealed under a layer of metal or large head, galvanized steel nails under felt stripping. Avoid exposed nails wherever possible.
11. In windy areas or on high-rise buildings, use stainless steel, Monel Metal, terne-plated stainless, or carbon-bearing steel flashings with stainless steel fastenings. Soft metals are not satisfactory.
12. At open roof edges, use a separate cant strip counter flashing rather than a stripped-in gravel stop. If a 6-in.-high cant is used, some roofs can be drained through scuppers to head boxes instead of using a full-length outside metal gutter.
13. In poured concrete, concrete block, and brick walls, use flashing reglets (metal or PVC) embedded in the wall or mortar joint and flashed with a spring-lock type of counter flashing, wedged, caulked, or closed with preformed compressible strips.
14. Cover precast concrete parapet walls with metal after caulking vertical joints inside and outside with suitable sealant. Concrete slabs require wood nailing strips or blocks for attaching metal.
15. Wood nailing strips and blocks should be made from dry lumber, pressure treated with waterborne preservatives, shaped so that they will not fall out of position, and be a wood species that holds nails and screws. Brush treatments are not recommended.

16. Cant strips should be made from solid lumber as in item 15, secured to the deck only. Air spaces behind or under cants should be avoided.

17. Expansion joints or control joints are required in the roof only where structural joints occur or where the deck changes direction (e.g., a building shaped like an L, U, E, F, or H). All such joints should be raised above roof level with curbs. Flat expansion joints are not recommended.

18. Provision may be required for movement at the ends of long-span, double-tee, precast concrete sections when changes in moisture and temperature are possible in the concrete or when rolling loads (cars) cause deflection.

19. When window or door openings are adjacent to a roof, the bottom of the sill should be not less than 6 in. above the roof. Roof water should be diverted away from such openings. Protection should be provided for the roof outside doors to prevent mechanical damage from foot traffic.

20. Flashings around projections through the roof should allow for movement by using separate base and counter flashings.

21. Avoid the use of gum pans or pitch pockets unless the roof membrane is continuous under the pan. Use hot type 1 asphalt for filling gum pans if they cannot be avoided. Solvent-type gum is not recommended.

22. A gummed flashing against a wall requires a caulking material that will adhere to the metal flashing and the wall material without shrinking or deterioration due to weather exposure. Cleaning and priming may be necessary. The metal flashing should be made from a metal with a low coefficient of expansion and held in place with bolts or screws through slotted holes in $1/4$-in. by $1 1/2$-in. bar steel or $1 1/2$-in., 16-gauge channel, preferably galvanized. Slotted holes allow thermal expansion of the bar without buckling. This is not as good as a reglet flashing and should only be used where there is no alternative.

23. The counter flashing at walls where the wall cladding overlaps the flashing should be made in two parts so that the lower part can be removed when necessary without disturbing the upper portion or the cladding.

24. Joints between metal sheets should allow for thermal movement. (See Figure 13.1 regarding thermal expansion of metals.) An increase in metal gauge and more frequent fastening, or the use of a continuous clip or cleat, helps to reduce rippling or oil-canning. Jointing details are shown in the flashing drawings.

25. Porous walls above flashings must be waterproofed or water may run behind the flashing and into the roofing system. Low parapet walls of brick and block or other porous material should always be covered with metal. Under extreme conditions these walls should be covered on the outside with a rain screen.

26. Metal flashing details in architectural drawings should be drawn at a scale of one-quarter full size for clarity. This scale eliminates any doubt as to what is required.

27. Figures 13.2 through 13.18 illustrate some of the basic principles of flashing built-up roofs. There are many acceptable variations of their application to satisfy a wide range of structural, geographic, and environmental differences.

Thermal Expansion of Metals and Flashing Details

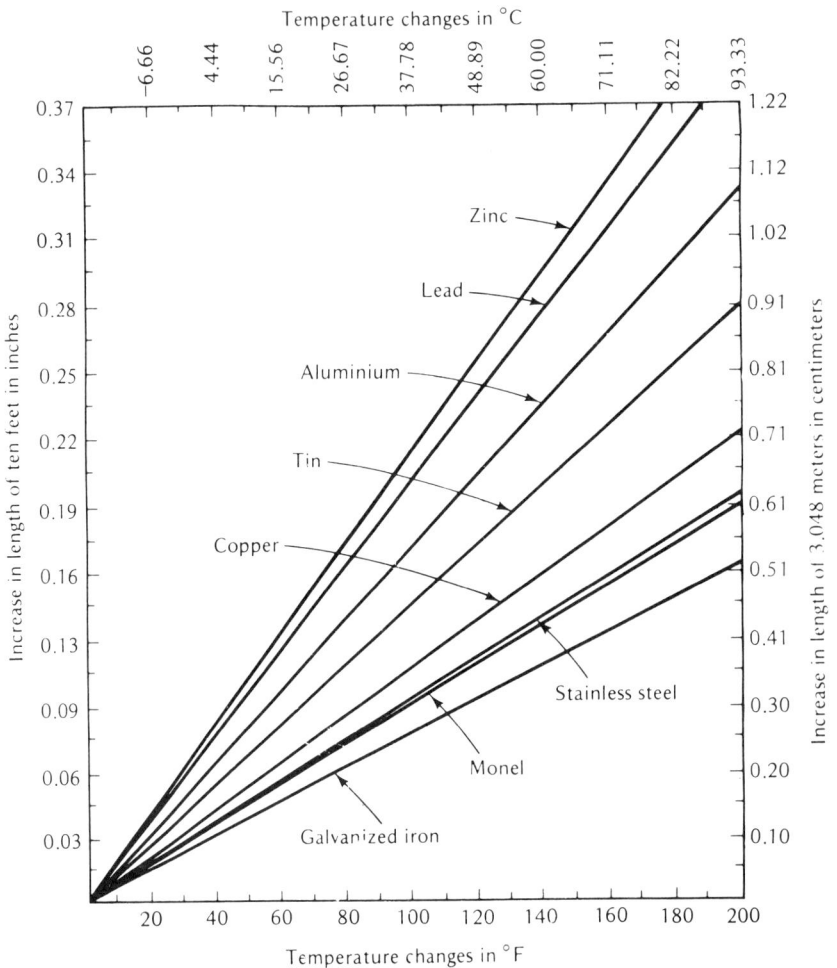

Figure 13.1 Thermal expansion of metals. (Reprinted by permission from M.C. Baker, "Flashings for Membrane Roofing," Canadian Building Digest No. 69, Sept. 1965, DBR/NRC Ottawa, Ontario.)

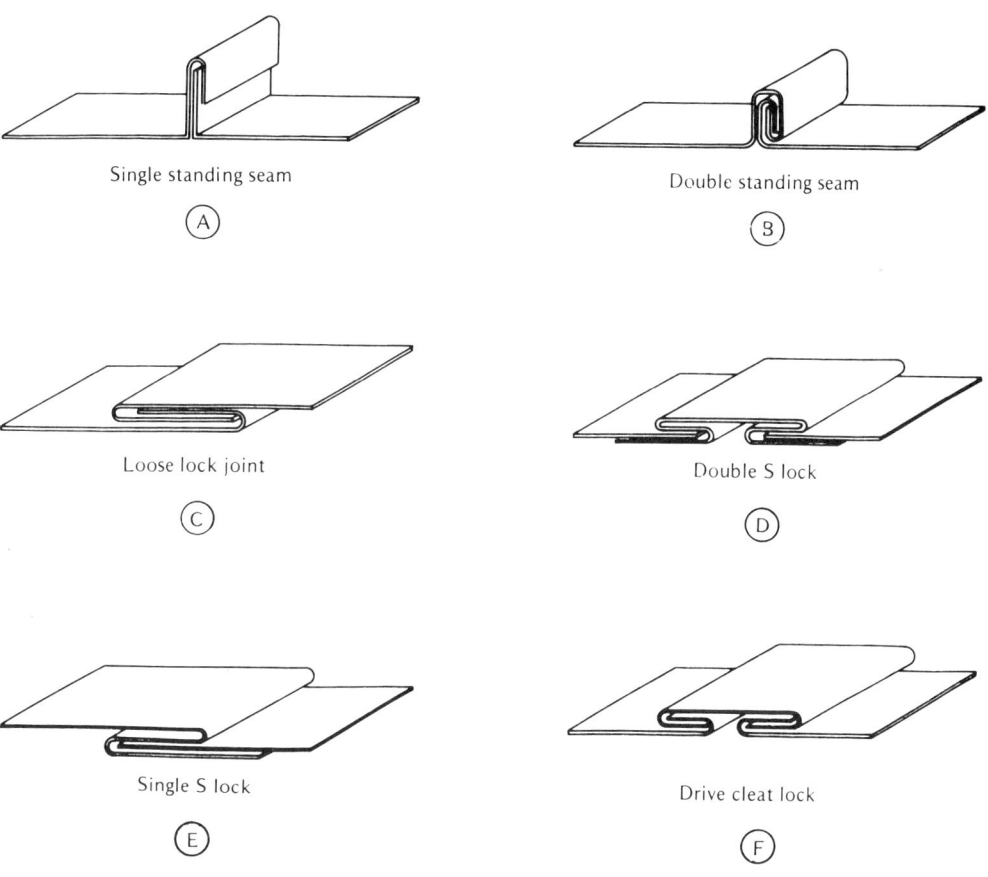

Figure 13.2 Flashing joints.

133

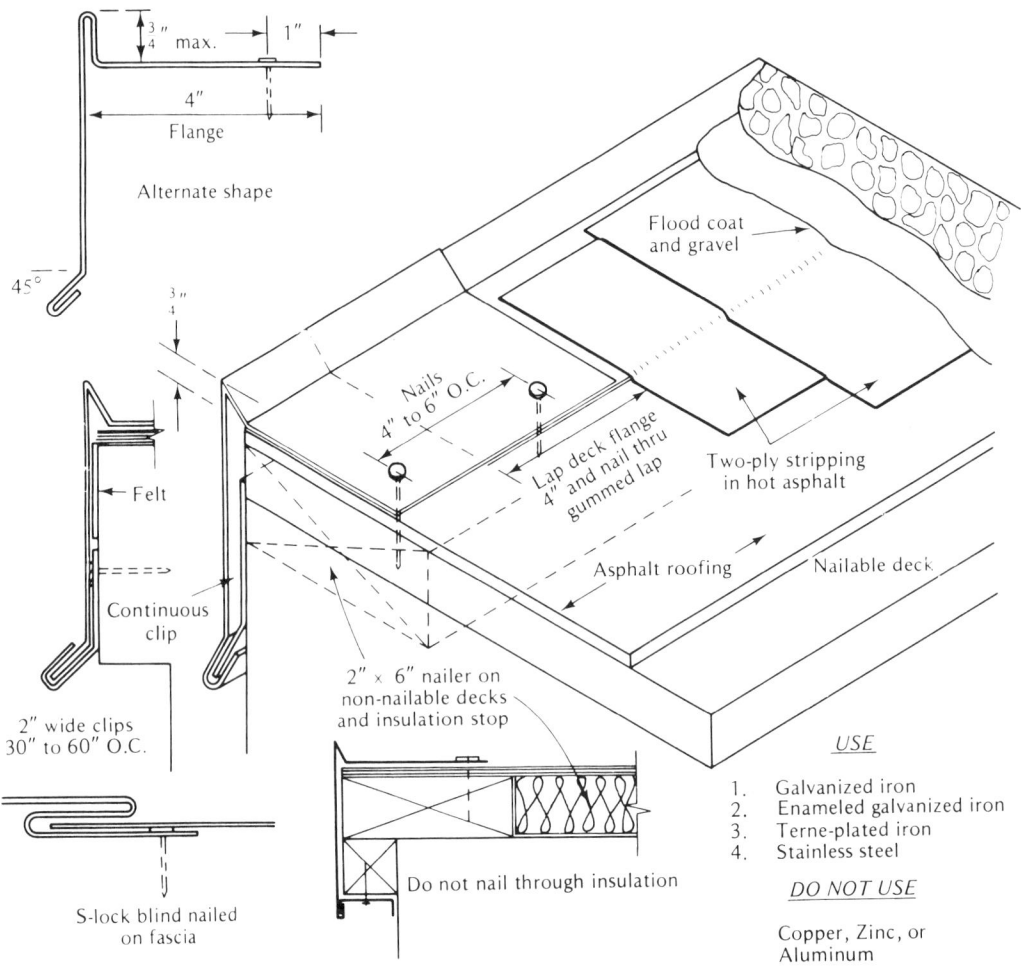

Figure 13.3 Gravel stops.

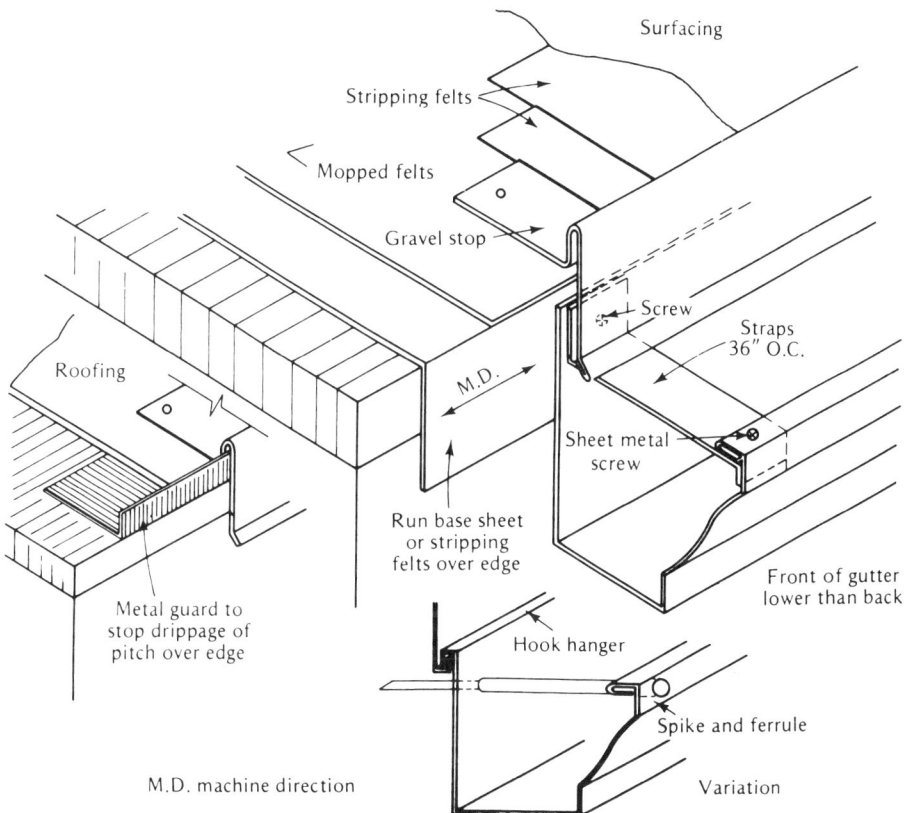

Figure 13.4 Gravel stop and gutter.

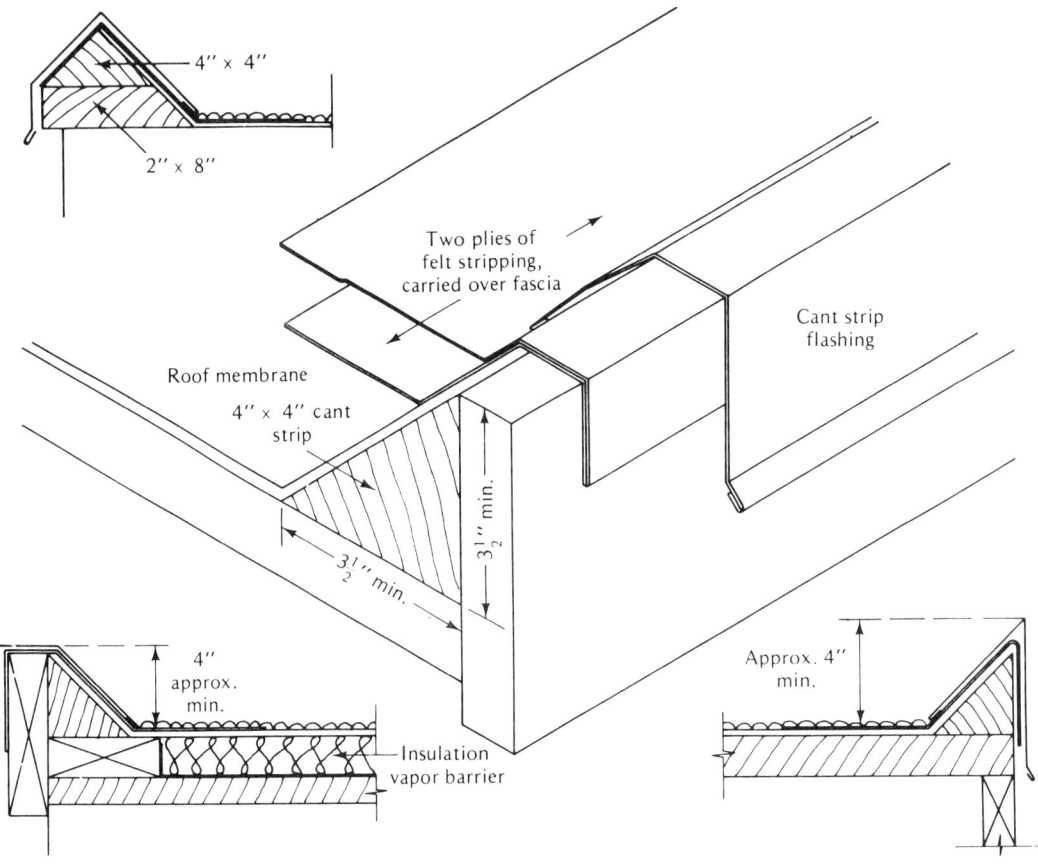

Figure 13.5 Cant strip at roof edge. (See Figure 13.6 for reverse view.)

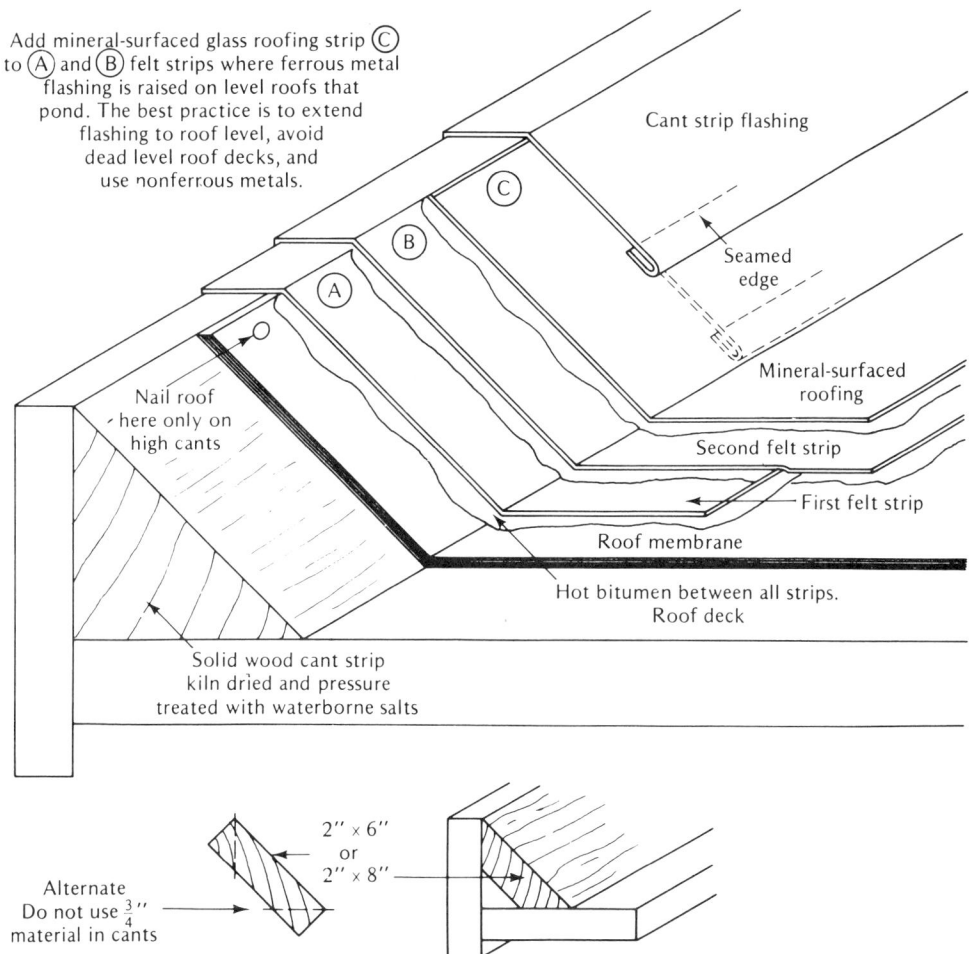

Figure 13.6 Cant strip at roof edge (reverse view).

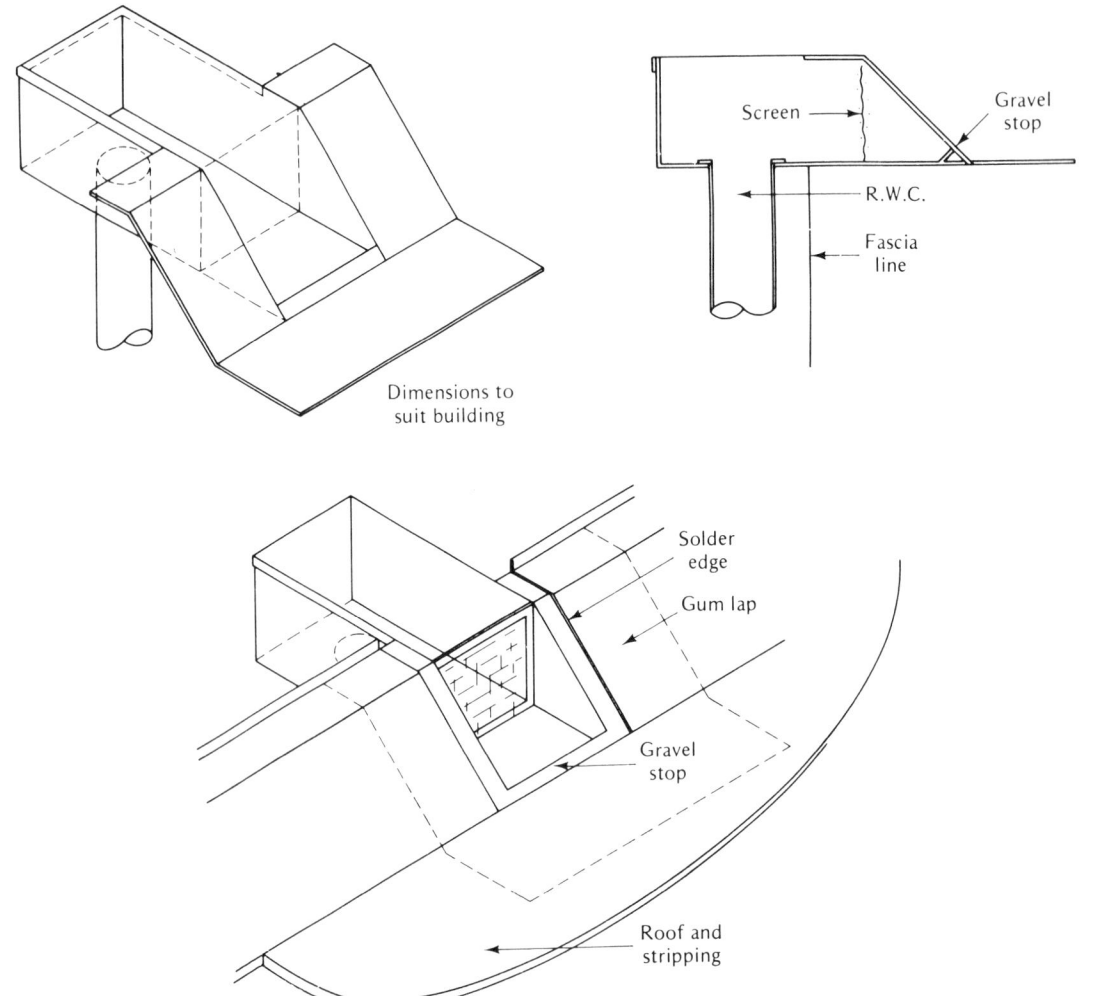

Figure 13.7 Scupper drain through cant strip.

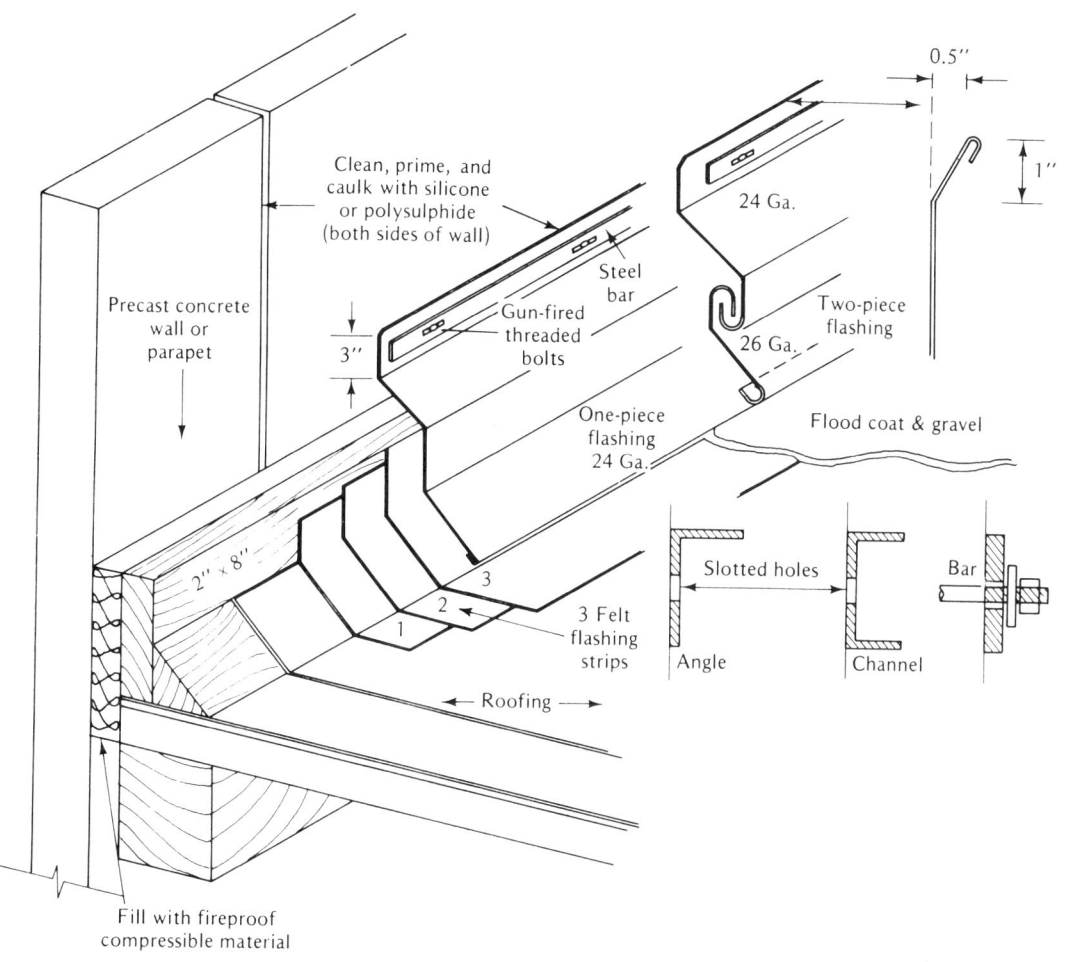

Figure 13.8 Caulked wall flashing.

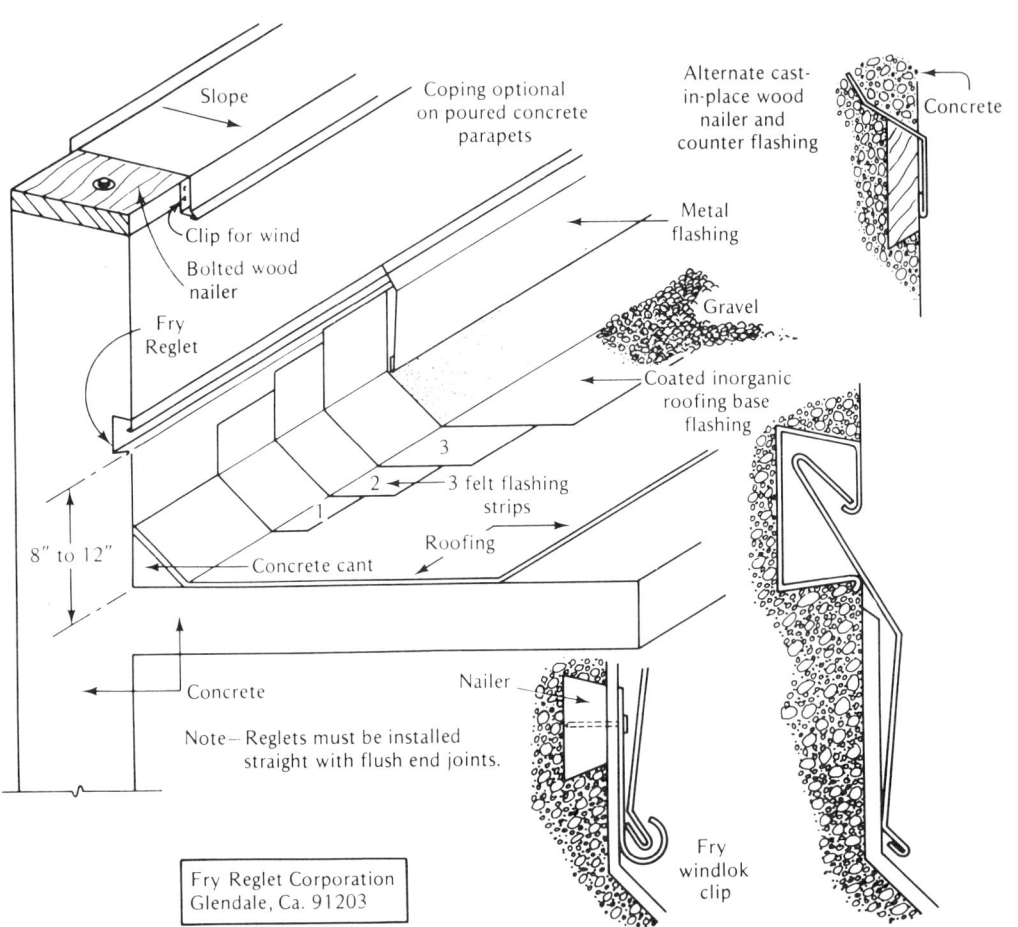

Figure 13.9 Flashing poured concrete wall.

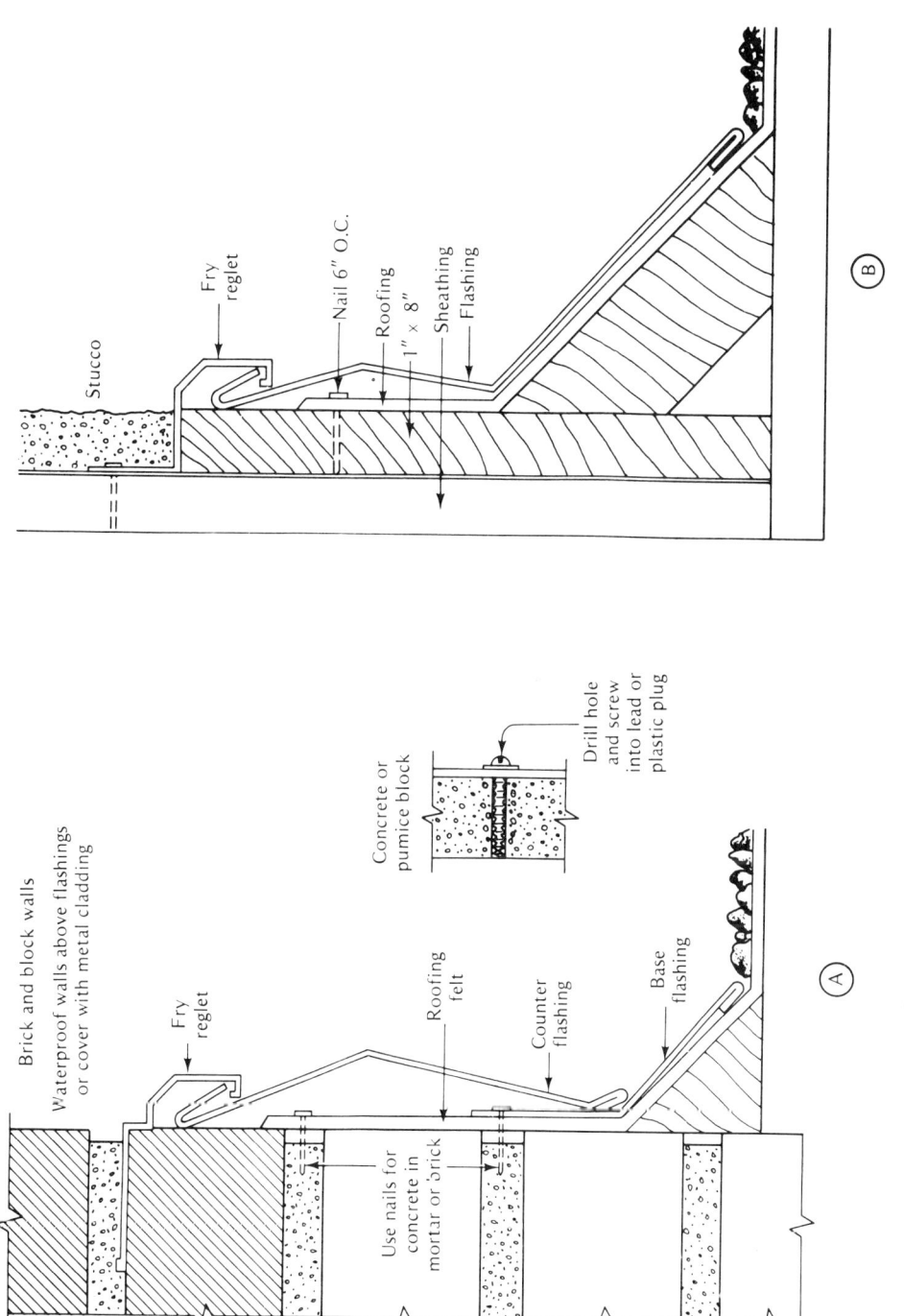

Figure 13.10 Flashing at walls: (A) Brick and block. (B) Stucco.

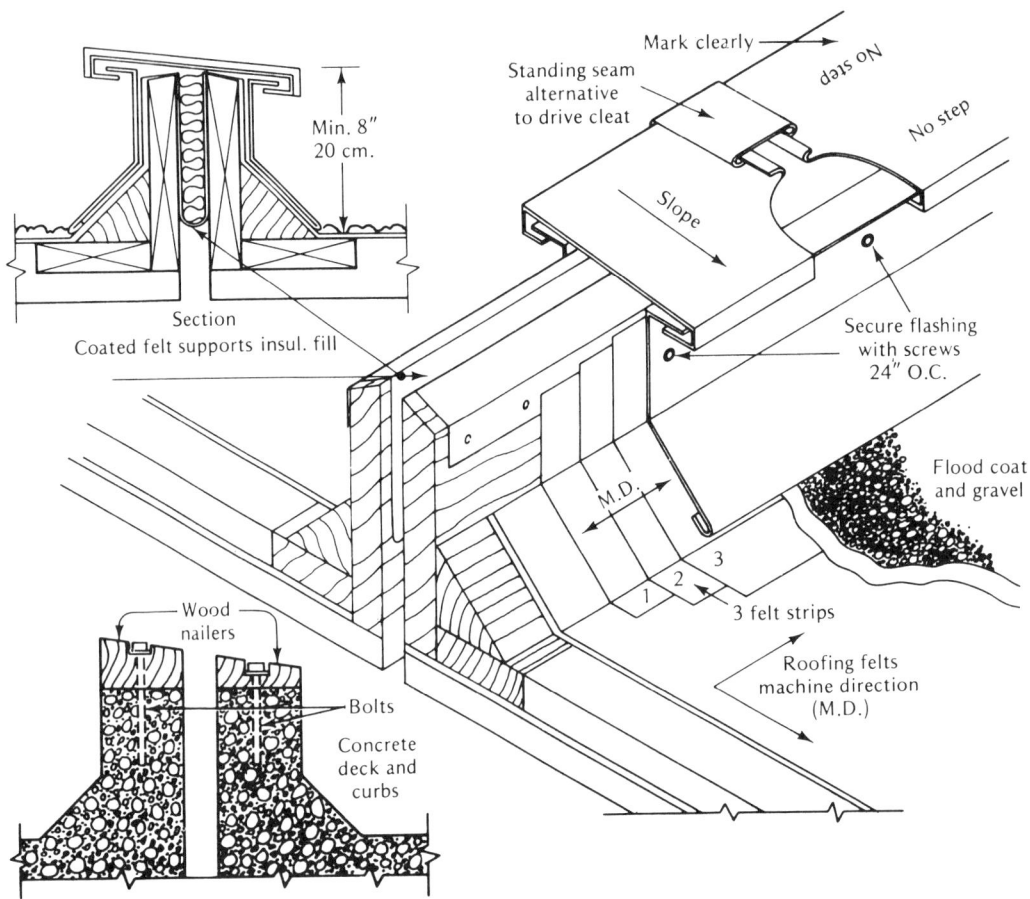

Figure 13.11 Expansion joints.

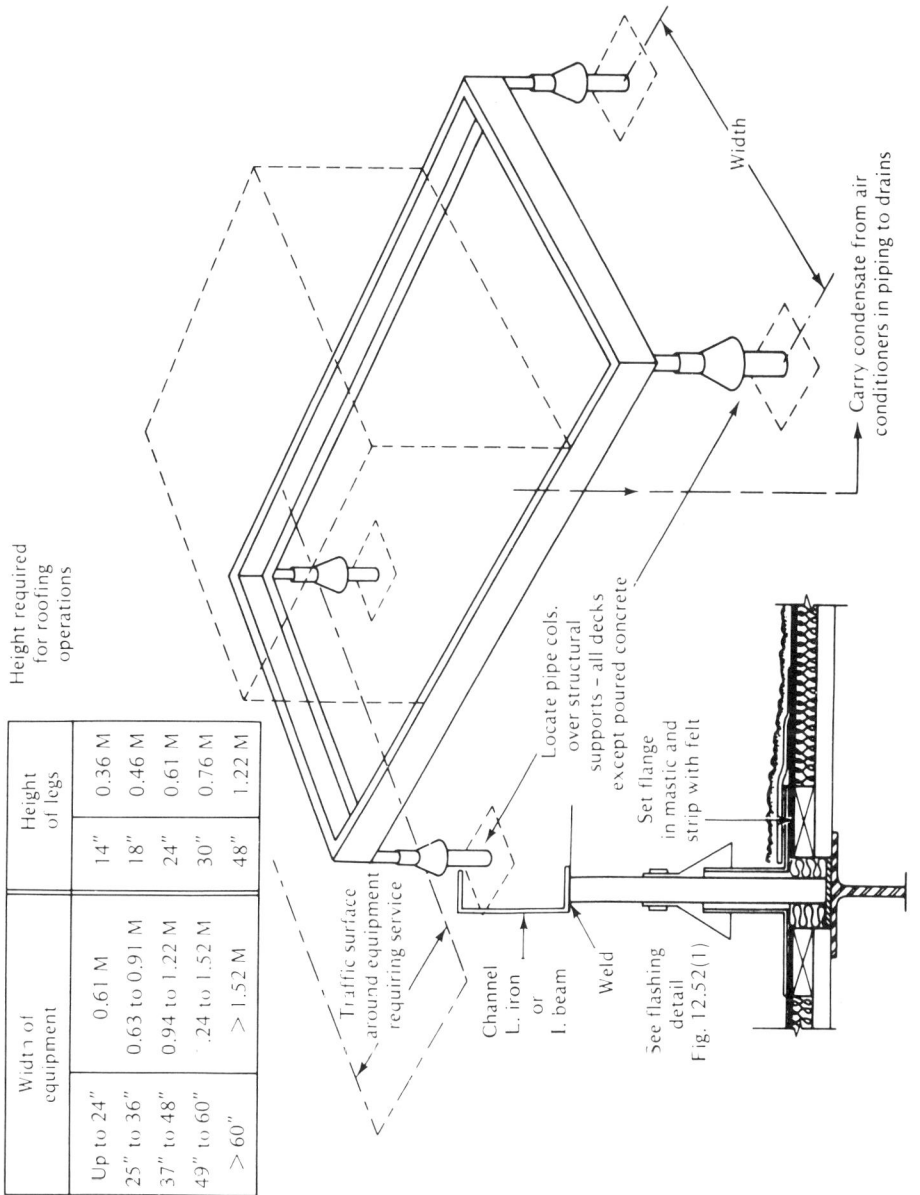

Figure 13.12 Mechanical equipment stand. (Based on a design by N.R.C.A., Chicago Ill. 60631.)

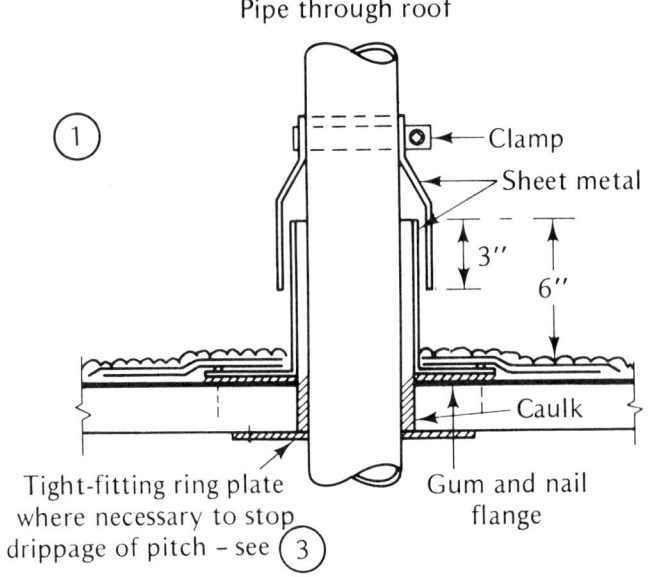

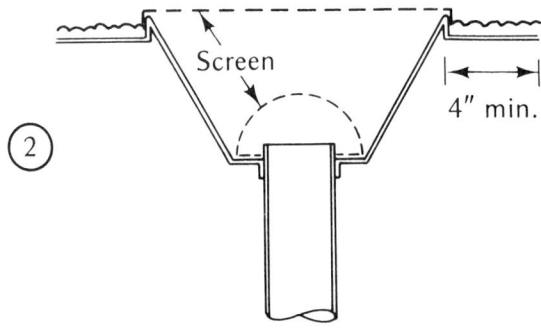

Figure 13.13 Miscellaneous: (1) Pipe though roof. (2) Sheet metal drain sump.

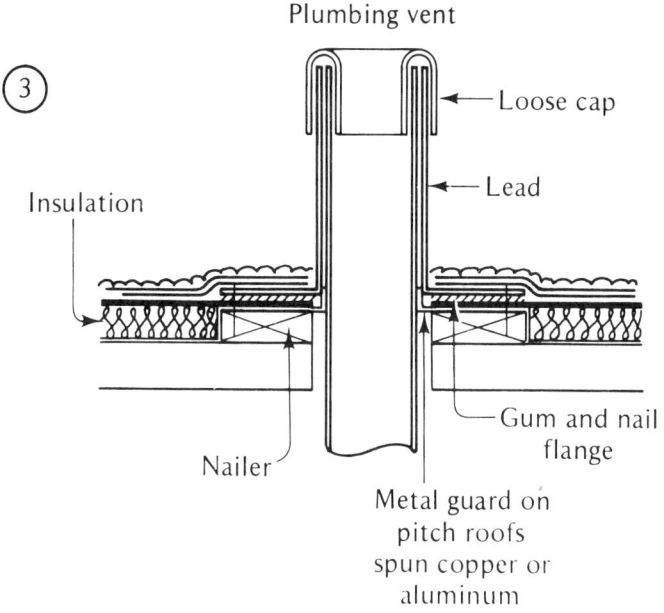

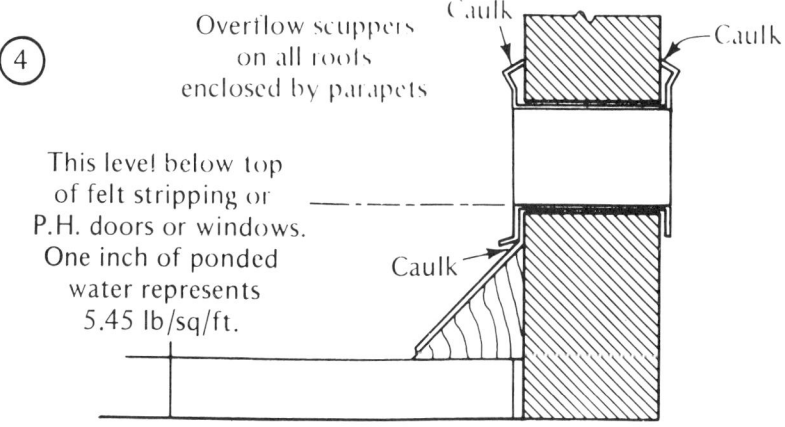

Figure 13.14 Miscellaneous: (3) Plumbing vent. (4) Overflow scupper.

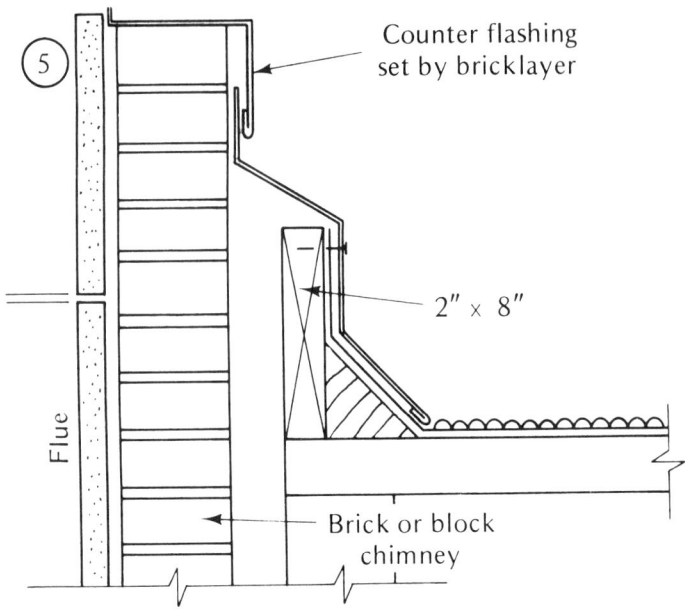

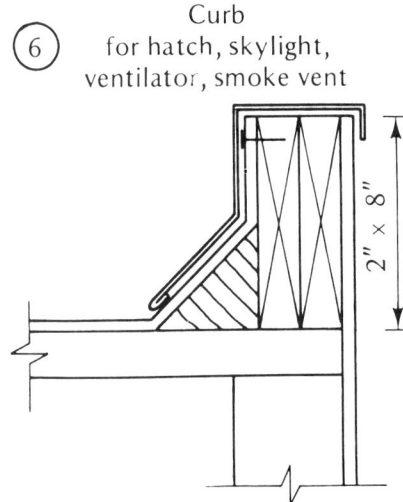

Figure 13.15 Miscellaneous: (5) Chimney flashing. (6) Curb flashing.

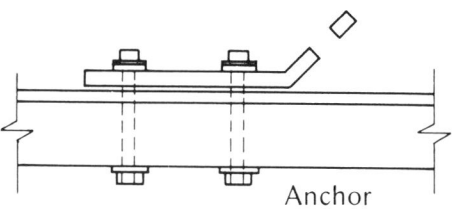

Anchor

NOTES

1. Roofing must be continuous under all fasteners.

2. Iron, as well as bolts and washers, etc., must be galvanized.

3. Neoprene washers under all steel washers.

4. All iron bases set in plastic gum before gravelling.

5. Size sufficient to support weight or stress on bases and fasteners.

6. All installations on solid bearing – i.e., not through insulation.

7. None are recommended on ponded roofs.

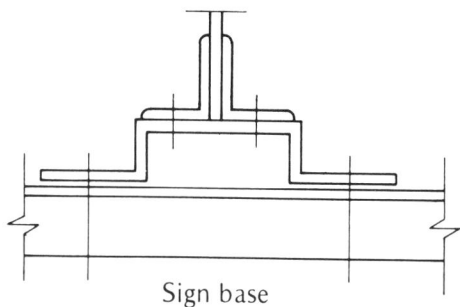

Sign base

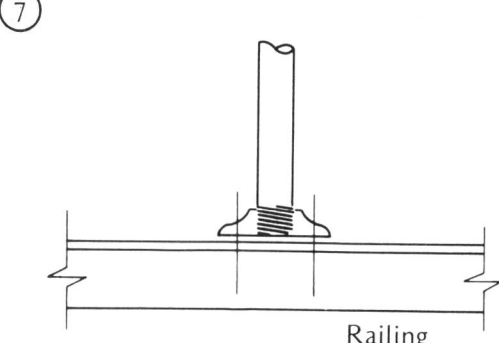

Railing

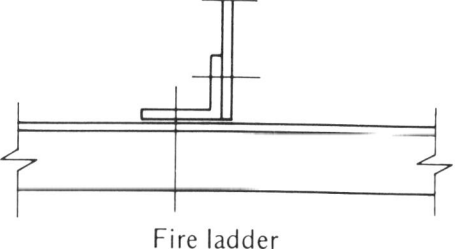

Fire ladder

Figure 13.16 Miscellaneous: (7) Anchor, sign base, railing, fire ladder.

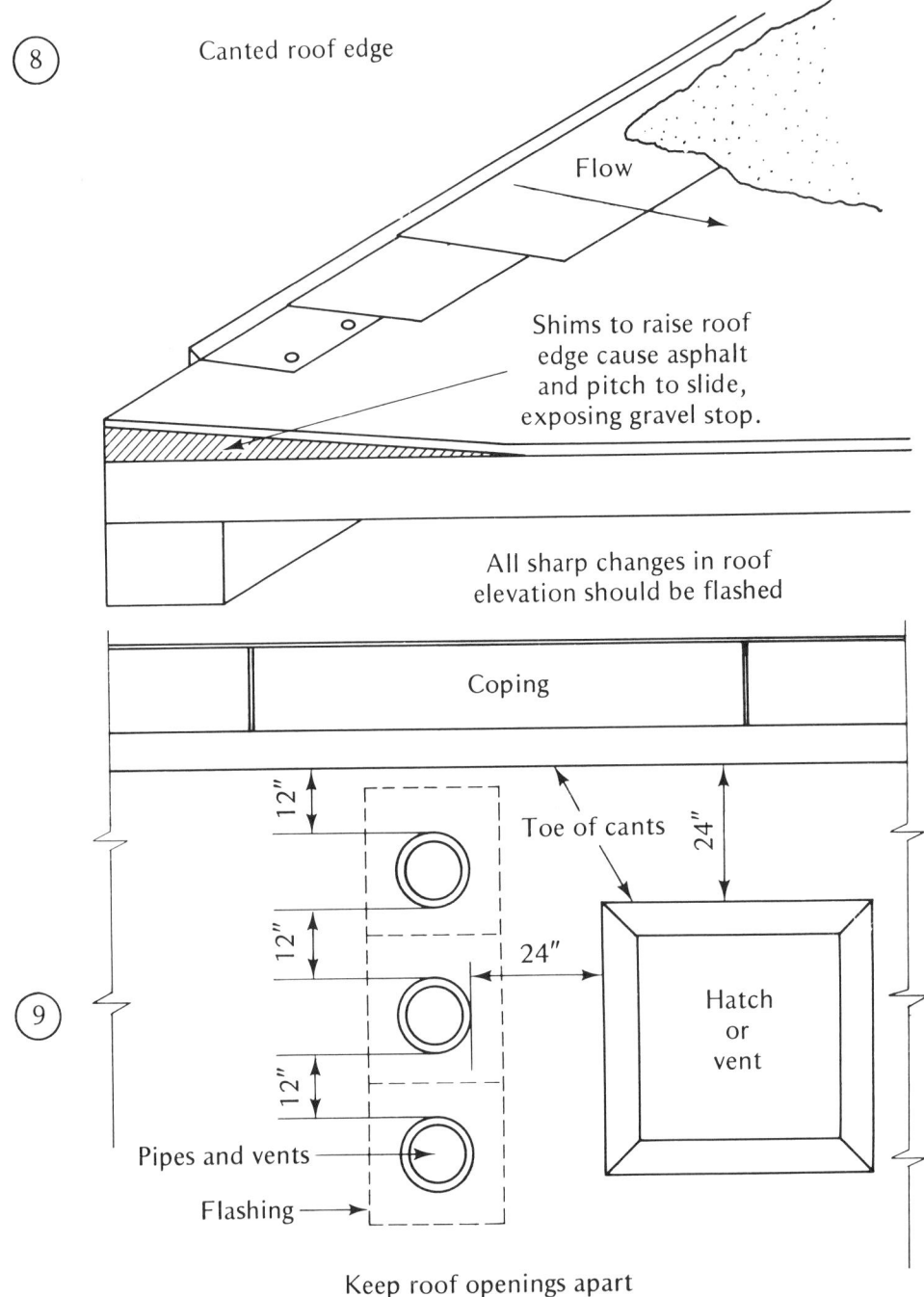

Figure 13.17 Miscellaneous: (8) Canted roof edge. (9) Roof openings.

Walks and patios

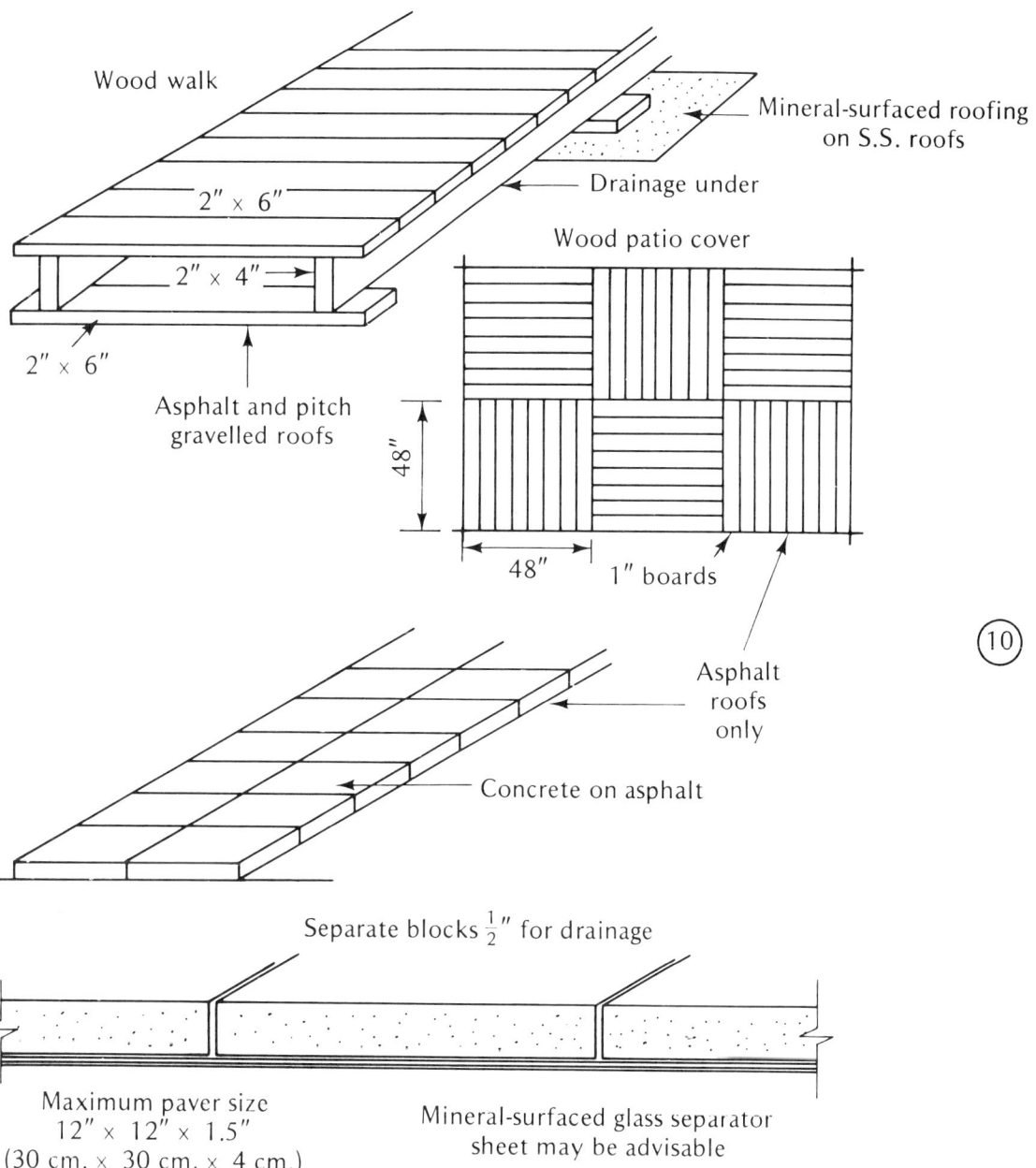

Figure 13.18 Miscellaneous: (10) Walks and patios.

14

Application Procedures and Workmanship

Hot built-up roof membranes are site assembled from several components into a continuous or monolithic unit. Although a considerable amount of mechanical equipment is employed, the final result is not always consistent within narrow limits or tolerances. In view of the many variables that exist, some variation should be accepted. However, the roofing system must remain watertight for a reasonable number of years.

Heating Bitumen

Asphalt and coal-tar pitch in paperboard cartons and metal drums are broken up while in the cold state and heated in liquid propane gas or kerosene-fired kettles of several designs and sizes, from a 25-gal patching kettle to a four-wheel, 600-gal trailer type weighing 2,300 lb. Bitumen is also delivered to the job site in heated tanker trucks in which the bitumen is raised to the optimum temperature and pumped directly to the roof or into storage tanks. This equipment is generally reserved for large roofing jobs.

All storage and heating devices are equipped with thermometers, but they do not always remain accurate over a long period. Dial dip thermometers and armored mercury thermometers registering up to 650°F are used by roofers and inspectors to check temperatures in the kettle and in the mop bucket. With the introduction of the

Application Procedures and Workmanship

equiviscous temperature guide, portable automatic control systems, which are available, will be required to control temperatures in kettles, tank trucks, and storage tanks more precisely.

Special two-wheel carts are made to receive the hot bitumen from the pump outlet at roof level. The flow is regulated by a person on the roof as much as 100 ft above the kettle. Ten- and twenty-gallon carts are used when hand mopping felts and 25-gal carts for flood coats. Other larger units are used to transfer hot bitumen to wheeled felt layers, which apply a mopping layer and a ply of felt at the same time.

Laying Felt

Roofing felt can be laid in several different ways, and the one selected will depend on the roofer's preference on each job. The general criteria is that felt be rolled out straight with the correct amount of lap and minimum curving or wandering off the guide lines. All air pockets must be broomed out and all voids filled with hot bitumen. The bitumen should extend out beyond the exposed edge of the felt and form a puddle in front of the roll. No felt should be rolled into cooled bitumen, which means the felt rolls should be embedded in the "hot stuff" (roofer's slang description) as quickly as possible. Specifications often call for "brooming in" the felts. This is done more or less automatically with steel chain links, which are part of the felt layer (see Figure 14.1). However, many roofs do not lend themselves readily to felt layers, which sometimes require two workers for safety. The operator walks backward. Hand-mopped No. 15 felt rolls weighing 60 lb seldom require brooming on smooth decks

Figure 14.1 (A) Felt layer deposits bitumen and rolls in felt as the operator walks backward. Felt should be broomed in by another worker to ensure good contact. (B) A similar type of felt layer with a chain being dragged over the felt instead of brooming. This is not particularly effective.

like plywood, except when the last quarter of the roll (15 lb) is being laid. Lightweight glass felts and all felts on rough uneven decks do require brooming in order to ensure contact at all points.

Problems arise with some felts in hot weather when type 1 asphalt or coal-tar pitch bleeds through the felt. In this case brooming becomes very difficult. It may be caused by overheating the bitumen so that the viscosity falls too low or because the felt is too porous.

Any felt (usually organic) that buckles or fishmouths at the edges or curves off line should be rejected. Felt rolls that have been flattened or damaged on the ends during shipment or storage may be impossible to lay owing to rippling and tearing. All felts should be handled carefully from the delivery truck to temporary ground storage and to roof level to avoid damage. Felts should always be stored on end on a dry surface and protected from the weather.

Roofers will occasionally install a vapor barrier, insulation, and one ply or felt on one day and follow with two or three plies the following day. If the first ply is not glazed over, the felt may absorb moisture overnight, which will be trapped under the felt laid in the second day. Felts are also laid two and two with the same result. When soft asphalts or pitch are used with organic felts, these practices often lead to curling of the felt edges in hot sun if there is any delay in applying the second group of felts. Once a wood-fiber organic felt has curled at the edges, it is generally impossible to flatten. It is not advisable to use a tilting wheeled cart to spread bitumen because the amount of flow cannot be controlled, and it cannot be deposited exactly where it will be covered by felt. The tendency is to run the cart out too far ahead of the worker who is rolling in the felt. This is not good roofing practice.

When equipment is used that deposits bitumen on the roof through adjustable openings in the bottom of the container the opening size and the viscosity of the bitumen must be compatible so that the correct amount is applied. If openings become clogged, there will be skips in the flow and a poor roof will result. Very often the skips will not be noticed at the time of application, but will show up later in the form of long blisters or buckles in the roof membrane.

Hand mopping may be slower, but the operation can be most easily monitored. The hot bitumen should flow easily off the mop, which is usually fiberglass from 4 to 7 lb each. When a mop wears down, it will not hold enough bitumen, and the tendency is for the mopper to scrub the bitumen into the roof instead of letting it flow on (see Figure 14.2).

Laying Base Sheets

Best results are obtained when coated base sheets are laid when the temperature is above 50°F (10°C), and after the material has been rolled out flat for a few hours to relieve the tension in the roll. Some uncoated asbestos sheets and coated glass may be easier to lay in cool weather.

Depending on the weight, base sheets are available in 108-ft and 216-ft rolls to cover exactly one or two squares when lapped 2 in. They may be mopped directly to the deck or substrate, or nailed through discs at laps and approximately 12 in. on center in both directions. Stapling through light-gauge aluminum or steel discs may be permitted in some areas. Stapling, nailing, or mopping is governed by the nature and holding properties of the substrate and the degree of wind suction expected.

Figure 14.2 (A) Hand mopping and gentle rolling in of felt. Mop is too small and mop bucket should not sit on roof surface. (B) Felt being rolled into a puddle of asphalt in order to fill all voids between felt plies. The old roofing is being stripped back at the left.

Asphalt base sheets are not recommended for tarred felt roofs.

Coated base sheets may reduce but will not eliminate the wrinkle cracking failure of roofs laid over thermal insulation and a vapor barrier.

Base sheets are nailed, never mopped solid, to roof decks containing water (e.g., decks 5 and 9 through 12).

Inorganic materials are preferred to organic because of the lesser tendency to absorb moisture and change dimension. Being heavier per 100 ft^2 than ply sheets does not necessarily mean that they have increased tensile strength, because the basis felt weights are often the same. At temperatures below about 32°F (0°C) there might be some improvement if there is coating asphalt on both sides.

Stripping Felts

On all roofs it is important that the edges be stripped on top of the roofing felts with at least two extra plies of felt of the same kind as in the roof membrane, both laid in hot bitumen. At cant strips these are often covered with mineral-surfaced roofing or an asbestos-cotton laminate to form a base flashing. Stripping felts are required to cover the nailed flanges of gravel stops, drain flashings, plumbing vents, sheet-metal vents, and other flashings of projections through the roof.

A simple slitting device, hand operated or motor driven, can be purchased or made by a roofing and sheet-metal contractor to slit rolls of felt to any width desired. This is preferable to cutting tag ends of rolls on the job with a roofing knife. This practice produces strips of uneven width and ragged edges and an unprofessional job. The widths will vary depending on the requirements of the job. Starting with the narrowest width, the stripping felts should be shingle mopped out on top of the roofing felts so that only the edge of the last and widest strip is seen.

Where type 1 asphalt or coal-tar pitch is being used, it may be advisable to install an extra ply or strip of felt on the deck at open edges and fold it back over all mopped felts to enclose and prevent drippage in very hot weather. This is done before sheet-metal flashings are nailed through the roof and stripped with extra felt.

Flood Coat and Gravel

As soon as possible after the roofing felts are laid and the edges and flashings stripped in, the roof should be poured with a heavy flood coat (60 lb of asphalt, 75 lb of pitch) into which the gravel or slag is embedded while the flood coat is still hot. These operations can be done by hand (Figure 14.3A) or with wheeled containers of bitumens (Figure 14.3B) and gravel (Figure 14.3C). The wheeled carts are preferred because they can be loaded directly from the hoist or kettle pump and transferred to the roof surface without delay and with minimum cooling. There is no need to pile wet gravel on the roof and then shovel it by hand into the "hot stuff." It is impractical in many cases to gravel in portions of the roof laid the same day unless several crews are working on a large roof. Generally, the felts are glazed with a light mop of hot bitumen, with the flood coat and gravel following when there is a sufficient area available to make the use of wheeled carts practical (Figure 14.3D). The possibility of curled felt edges must be considered.

Mopped Surface Coats

On steep roofs the surface may be a 30-lb mopped coat of type 3 or 4 asphalt applied as the roof is laid. It is generally advisable to wait one year before applying a reflective coating, which has a vehicle compatible with the asphalt. Mopped surface coats must be applied evenly or in a constant thickness. Thin areas will weather off, and thick coatings of high-softening point asphalt tend to alligator more readily. An alternative in warm weather is to apply a thin mopped coat of hot steep asphalt, followed by 2 to 4 gal of clay-stabilized asphalt emulsion per square. The extra reflective coating is still recommended, and it can be applied as soon as the emulsion is cured or dry. The curing of asphalt emulsion is simply the evaporation of its water content (approximately 45%). The rate of curing depends on the ambient temperature and relative humidity.

Vapor Barriers and Insulation

Vapor barriers, which also serve as air barriers, may consist of a factory-coated base sheet type of material with side laps mopped together, two or three plies of mopped No. 15 felt, or, in the case of a steel deck, a PVC sheet laid in chlorinated rubber adhesive parallel with the flutes in the deck.

There is no agreement in the industry as to whether it is advisable to seal off the insulation at all openings in the deck by extending the vapor barrier beyond the edge and back on the top surface of the insulation. Likewise, at the perimeter some authorities suggest that the insulation be butted against a wood strip and the vapor barrier be returned back on top of the insulation. The argument is that in the event of a flashing leak the insulation will be protected. This may be true on some decks, but would not hold for steel decks, for instance.

Other authorities maintain it is better to allow any stray air and vapor to escape at deck openings and roof perimeter before it condenses in the system. Roof system designers must decide for themselves which method they prefer, or whether to go the

Figure 14.3 (A) Spreading gravel by hand. (B) Spreading flood coat from wheeled cart. The operator walks backward. (C) Wheeled gravel spreader.

Figure 14.3 (D) Asbestos felt receiving a glaze coat of asphalt from a wheeled mop bucket so that all flood coating and graveling can be done at one time. Each job has a different timetable. The white lines indicate the roof is laid three ply.

route of the protected membrane system, where the waterproof membrane becomes the vapor barrier and the insulation is laid on the top or cold side.

In the vapor barrier-insulation-roof membrane sandwich system, the insulation is laid in hot bitumen when the melt point temperature of the insulation allows it. The insulation can be laid in multiple layers with asphalt or pitch in between to achieve whatever thermal resistance is desired. Although nailing of insulation should be avoided, it is occasionally necessary. It is suggested in multiple-layer construction that the first be nailed and the second mopped. Sliding of the second layer on sloped roofs can be prevented by ledger strips, high-softening-point asphalt, and heat-reflecting surfaces. The ledger strips are mechanically fastened to the deck through the first layer of insulation, which is continuous.

Insulation boards or sheets should be laid tight against each other to reduce damage at the perimeter caused by a gradual shrinkage of some roof membranes by thermal cycling. Open jointing is also dangerous, because it provides voids where moisture vapor can accumulate. The heating and cooling of an aqueous vapor under daily changes in temperature create pressures on the roof membrane contributing to wrinkle cracking failures over the joints.

Mineral-surfaced Cap Sheets

It is standard practice to lay two plies of mineral-surfaced 19-in. selvage roofing over mopped felts on roof inclines above 2 in. per foot. The selvage edge must be nailed

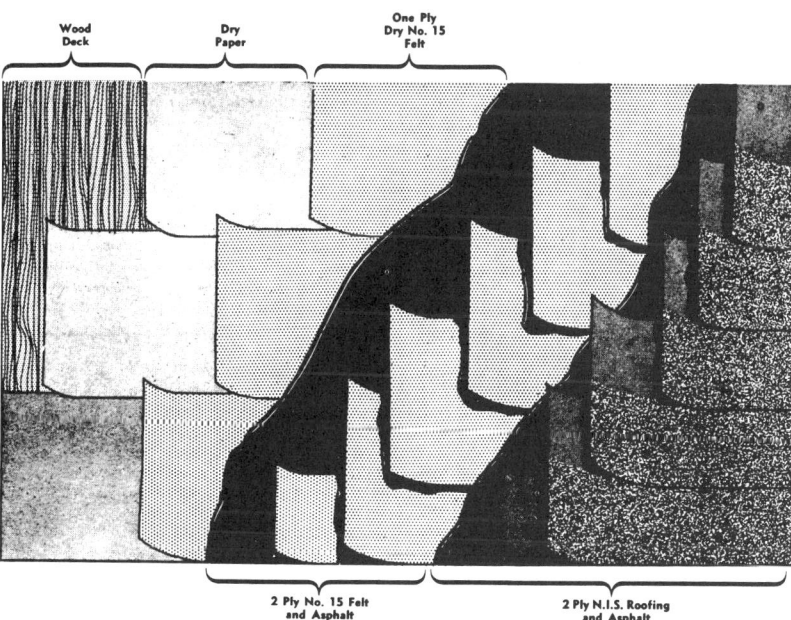

Figure 14.4 One method or specification for mineral-surfaced, split-sheet, mopped-in asphalt on a wood deck. Nailing not shown.

Application Procedures and Workmanship

through the mopped felts into the deck to prevent sliding, and mopped with steep asphalt to bond the plies together. Only nailable decks without insulation should be considered for this type of roof covering because of the widely scattered nailing points. An inorganic felt base in the selvage roofing is preferred to an organic felt owing to the lesser shrinkage across the width. Because of the difficulty in achieving complete asphalt coverage between plies on sloped roofs, there is a tendancy for mineral-surfaced cap sheets to blister and buckle. A black mineral-surfaced sheet may blister quite badly owing to high surface temperatures in summer (see Figure 14.4).

Figure 14.5 shows the taking of a cutout sample of a roof from the concrete deck. The 4 in. × 36 in. (one square foot) sample is returned to the roof and covered with four plies of felt shingle mopped out on to the roof. The purpose of taking the sample is to check the weight with the specification. Such sampling is not favored by most authorities since the integrity of the membrane is destroyed. If the sample proves to be underweight, the proper course of action to protect the owner's interests is not easy to decide.

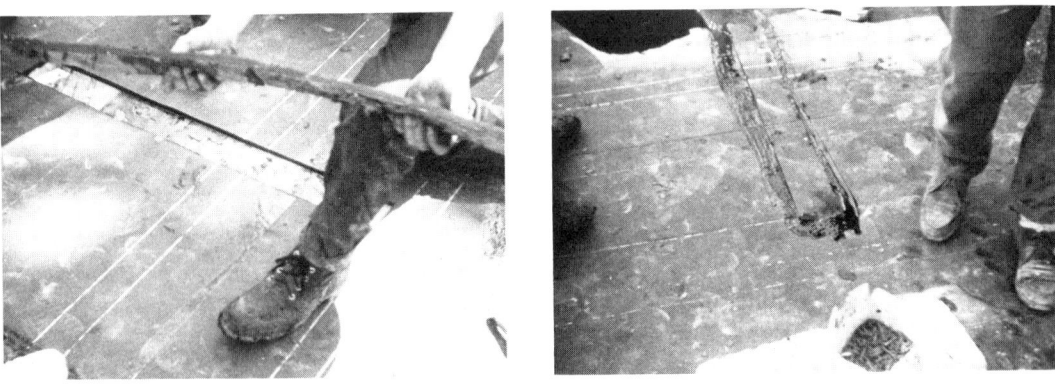

Figure 14.5 Taking a cutout sample.

15
Materials Handling and Storage

One of the difficulties attached to the construction of hot built-up roofs is the prevention of moisture gain by hygroscopic materials during shipment and during the storage period on the job. Moisture is present in most regions in the form of rain, frost, snow, high relative humidity, and dew. Moisture can be picked up from the ground or from dead-level roofs if materials are not protected by raised platforms and vaporproof ground sheets.

When materials leave the factory, they generally contain less than 10% moisture, but before they are installed in a roof system they can easily pick up another 10% or more through natural equalization with ambient conditions or by exposure to liquid moisture. The natural conclusion is that the use of inorganic materials requires somewhat less cover or protection against moisture absorption, but adsorption is still possible. The difference between the two words is significant. Mechanical damage to roofing materials can cause problems in application; therefore, handling procedures are important. Rolls must not be damaged by being dropped on the ends, piled too high, or piled in a horizontal position. Some types of thermal insulation, even when paper wrapped, are fragile, and can have corners broken through careless handling off a truck and to the roof level by a mechanical hoist (see Figure 15.1). Tank truck deliveries of bitumen ensure that the material is free of contamination, but package delivery to ground kettles often results in foreign materials being mixed in with the "hot stuff."

Figure 15.1 (A–C) Rolls of felt flattened and edges damaged by letting rolls lie flat. All roofers know they should be stored on end. (D) Coated glass felt severely damaged by the method of hoisting materials to the roof. This hoist flung rolls into the air.

If these solid materials reach the spigots in a felt layer or asphalt spreader, the outlets are blocked, causing skips in the mopping. Kettlemen should prepare a flat, clean work area to prevent contamination of the bitumen.

When gravel is delivered to the job, it is often dumped on the ground and then shoveled into a ladder hoist. Sand and dirt are sometimes scooped up and included with the gravel, which reduces the embedment and coverage in the flood coat. Transfer directly from clean storage to the hoist is advisable. The requirement of many specifications for dry gravel is not realistic. It should be clean or washed by the gravel supplier to remove dust and fines and can be delivered wet. The surface moisture is quickly driven off by contact with the hot flood coat and will stick. Adhesion will be somewhat better with types 1, 2, and 3 asphalt than with pitch; however, at low temperatures close to or below freezing, the gravel should be dry, free of frost, and preferably heated prior to application for both pitch and asphalt.

Asphalt in paperboard cartons may have piling instructions printed on the carton. If not, they should all be stored upright and not more than one tier high for types 1 and 2, and two tiers high for types 3 and 4, with plywood between the tiers. Heat and pressure combined with rain will collapse the cartons so that preparation for the kettle is much too difficult. The paperboard will be mixed up with the asphalt.

Materials Handling and Storage

Coal-tar pitch in metal drums must always be stored upright or the pitch will flow out of the top of the container, or through small holes punched in the sides, even when cold. Coal-tar pitch is better handled in tank trucks from which it will be pumped to the roof, eliminating a great deal of direct handling and ground-level heating, which is objectionable not only to the kettleman but to everyone else in the vicinity.

16

Reroofing Procedures

A built-up roof can be nailed to all roof decks except deck 7, poured concrete; deck 8, precast concrete; and deck 14, asbestos cement cavity. It should not be nailed to deck 15, insulation on structural deck, but sometimes this is done.

Nailing permits the removal of an old roof membrane without serious damage to or complete destruction of the deck. If self-clinching nails and caps have been used, as in poured or precast gypsum, the removal is not as easy.

When an asphalt roof is mopped to plywood decks 2 and 3, the plywood may be damaged when the roof is removed, but this will depend on the adhesive characteristics of the asphalt, the age of the roof, and the temperature at the time of removal.

A roof can be removed from poured concrete, deck 7, without damage to the deck. Depending on density, some damage could be expected with deck 10, lightweight concrete.

No roof mopped to an insulation base can be removed without seriously damaging the insulation as well as the vapor barrier under it. Roofs that require major repairs are generally in such poor condition that an insulation substrate will have been damaged because of leakage. It is also possible that the insulation failed because of moisture from inside the building, even before the roof membrane failed. The important point is that the two are so interdependent that when one fails the other

Reroofing Procedures

is rendered useless. This is not the case when insulation is below the roof deck or above the roof membrane.

When a roof membrane and insulation substrate are replaced, one has the opportunity to reverse their positions as suggested in the protected membrane system.

When an asphalt roof has been laid directly on the deck, it is sometimes possible to broom or scrape the surface to remove all loose and weathered material, and by priming and resurfacing to extend the life of the roof. This is suggested for asphalt felts because they remain soft and pliable for at least 25 years. It is not recommended for tarred felts more than 10 or 15 years old, because they may be hard and brittle and easily damaged by a mechanical scraper or broom.

No resurfacing is suggested for a roof that is sitting on wet insulation or one that has deteriorated by reason of the insulation.

Before reroofing or resurfacing is undertaken, the reason for the deterioration should be determined by a knowledgeable analysis of the system. This is an area where unscrupulous companies or individuals can take advantage of an owner's lack of knowledge by selling very little for a great deal.

It is not advisable to apply new felt over old because it is usually impossible to obtain intimate contact and a perfect bond. Air pockets between the old and new felts, plus moisture, could expand into blisters. If for any reason new felt is applied, it should be the same type as the old felt and as flexible as possible.

Old roofs are sometimes scraped smooth and covered with wood-fiber insulating board nailed through the old roof to nailable decks and mopped in the case of nonnailable decks. On nailable decks it is better to divide the insulation into two layers, the first nailed and the second mopped. A new roof membrane is mopped to the insulation. The old roof acts as a vapor barrier for the insulation.

The reroofing procedures described above should not be undertaken if the old roof has an insulating layer below it.

The Reroofing Dilemma

There are many opinions on flat roofing construction, expected performance, maintenance, and replacement procedures. In most cases, the opinions are supported by the desire to sell something.

A building manager faces a perplexing problem when a roof still leaks after a great deal of time and money has been spent trying to fix it. The company may not have budgeted for a major reroofing because of a depressed business cycle. One thing the manager will have is plenty of advice from people who have something to sell. The manager will also receive conflicting advice from people with their pet theories and pat answers to complicated problems. Somewhere in the middle there may be some valuable suggestions.

A few simple guidelines:

1. A leaking roof less than five years old is probably a bad roof. Unless condensation or a flashing leak can be proved, replace it.
2. If a roof is 15 or 20 years old, do not spend more than 20% of the cost of a new roof on repairs (approximately $20 per square).

The Reroofing Dilemma

3. If a roof has wet insulation under it, find out why before spending any money.
4. Discount offers of five- or ten-year guarantees on roof repairs and ignore the person who makes the offer. No one can guarantee a roof repair unless they know there is nothing wrong with the roof to begin with.
5. Make sure metal and fabric flashings are in good condition and are not leaking.
6. Check carefully for surface damage caused by roof traffic.
7. Find out as much as you can about the history of the roof and how the building has been used. The exact age of the roof is useful.
8. If a cut-out shows the original roof has been coated before or has had a second roof applied, replace the whole lot.
9. Apply coating or resurfacing to original roofs only when they are in good condition.
10. By examining cut-outs, identify all the components of the system and the number of plies or layers.
11. Check parapet walls for leaks, both inside and outside.
12. On dead-level decks that pond with water, see if additional drains are possible. On roofs enclosed by parapet walls or raised curbs, make sure there is an overflow scupper. Avoid controlled-flow drains.
13. Soft insulation like glass fiber or low-density plastic foam under a roof membrane may make it impractical to resurface the roof because of damage to the roofing fabric by high wheel loading and foot traffic.
14. The strength of the roof deck should be determined before any work is undertaken or additional dead load applied.

An owner may be faced with the dilemma of a leaking roof that has been patched for several years with no success, and the offer of a solution in the form of a new roof laid over the old. This can be tempting principally because it does not interfere too much with normal building functions and is much cheaper than a tear-off. Also, a minimum amount of old material is removed. Loose gravel and dust can be lifted off by heavy-duty vacuum equipment and piped directly into closed trucks. This is infinitely better than using power or hand brooms that produce vast amounts of dust without getting rid of it. If felts and bitumen and perhaps insulation are removed, high costs can result due to the transport to an approved dump site. There may be restrictions on how some materials are disposed of, e.g., toxic tarred felt and pitch, or nonbiodegradable plastic foams. Both of these materials cannot be burned without producing objectionable smoke and fumes and cannot be buried because of the long time it takes for them to disappear.

Continuous flat roofing membranes are not easy to fix once they begin to leak seriously. As a general rule, the damage is done from below and is not due to ordinary wear and tear of the elements or to sloppy workmanship. This means that recoating, resurfacing, or applying additional felt will not cure the basic problem, which is poor structural and roofing design and poor selection of roof system components.

Reroofing Procedures

A roof system that is not leaking, but that is showing signs of wear due only to weather exposure can be resurfaced quite economically and have its normal life extended. This is comparable to painting a piece of wood. Note, however, that the roof is resurfaced, not recovered.

A claim that an old roof can be resaturated should be looked at with a jaundiced eye. Old tarred felt is usually so brittle and weak that it cannot be made soft and pliable again except by some mysterious and as yet unknown alchemy. The rock-hard pitch coatings between the felts cannot be softened and cannot be penetrated. It is impossible to duplicate the conditions that existed in the roofing machine when the felt was originally impregnated with coal tar. Old asphalt felt roofs remain flexible much longer than tarred felt and can be primed and resurfaced with a cold cut-back primer and a flood coat of hot asphalt. This introduces a hot kettle, but many cold coatings have to be warmed slightly before they are applied. Some hazard exists in this operation because of the volatile solvents. The roof must be in good condition before resurfacing is attempted.

Buildings are always designed to support a predetermined dead load or a fixed unvarying load of roof deck, insulation, and roofing. They are also designed to support, without undue deflection, the positive live loads of ponded water, snow, ice, human traffic, and negative wind loads (suction). When new roofing materials are added to an existing roof, the margin for live loading is reduced. The dead load is increased and must not be so great that deflection is excessive, which might increase the ponding potential and stress the deck and supporting structure beyond acceptable limits. In addition to this loading there is also the temporary effect of heavy machinery to consider, e.g., spudders or scrapers, gravel buggies, vacuum equipment, and storage of materials.

As a general rule, it is not advisable to lay new roofing materials over old because blistering can be caused by air pockets and trapped moisture. A new roof should never be laid over an existing insulated system that is wet. Installing breather vents will not dry out the old system and may do more harm than good by helping the flow of air into the roof system from the interior.

The surface of an old roof can be prepared by removing all loose gravel and dusty materials with vacuum equipment, and then primed and resurfaced with hot bitumen pour coat and gravel. Do not re-use the old gravel and do not use only mechanical or hand brooming. Resurfacing an existing asphalt and gravel roof will require approximately one gallon of cold primer, 60 to 75 pounds of hot asphalt, and about $1/7$ cubic yard of new gravel. For roofs dead level to $1/2$ in. per foot, use Type I asphalt; $1/2$ to 1 in. per foot, use Type II asphalt, and from 1 to 3 in. per foot, use Type III.

Excessive amounts (7 to 10 gallons per square) of cold cut-back asphalt, coal-tar or emulsified asphalt and pitch coatings are not recommended for resaturating or coating old roofs because they are sold at inflated prices and do not perform as advertised. They leave less than 50% of their original weight in solid material on the roof after the solvents or water have evaporated. This is less than half the normal 60-lb flood coat of asphalt and one-third of a 75-lb flood coat of pitch per square. Generally, only a short time after one of these resaturating efforts takes place, it will be found that the roof has to be replaced after all. The extra sticky and oily materials that have been slopped around, plus the overabundance of gravel to make the roof look good, have to be removed along with the old roof. The removal costs are much greater than they would have been before the extra resaturating material was added.

Many building owners take the route of roof coating in the hope that they will save money and prolong the life of the roof just a year or two longer. Sometimes they think that a few patches and a coating will stop the leaks. Evidence abounds that the demand is there for roof coating and that it is being satisfied by at least 70 U.S. firms making roof coatings and cements, and by over 50 firms making aluminum coatings. Some of these companies do not wait for the business to come to them but adopt extremely aggressive sales techniques and advertise in the best business magazines to sell their products. They even offer a form of guarantee along with their pails and drums.

The author has been involved in the attempted restoration of smooth-surfaced asphalt roofs on industrial buildings where repeated coatings had achieved a thickness of 0.75 in. to 1.0 inch. These roofs, all sloping, were still cracking through to the wood decks, and no amount of patching was effective. In one case, the first roof was two layers of mineral-surfaced roofing (SIS or NIS), and in the other it was two layers of asbestos felt. These roofs were coated over and over again instead of removed and replaced. This practice could be compared to a lifeboat picking up so many survivors that it sinks, or painting yourself into a corner of a room.

If an old insulated system has failed, it should be replaced with one that has a better chance of surviving. A great deal will depend on the type, age, condition, and strength of the roof deck. If it is dead level and encloses a high-temperature and a high-humidity area, a new roof might not be an improvement. Admittedly, a new roof is expensive and sometimes difficult to accomplish without serious disruption to the interior; however, if a building has not been intelligently designed to allow for roof replacement, then one must accept the consequences. When it is known that roofs seldom last even 20 years and never more than 50, some foresight should be exercised by the designer and the owner in the design of the deck. It is often the owner who places financial restraints or limits on the architect, who must sacrifice cost and quality where possible. It is not unknown, and is becoming more common, for a built-up roof laid over insulation to fail in the first five years or less, and yet we keep on designing them the same old way, on decks stressed to their limit.

A reinforced poured concrete deck makes it easy to replace a roof. A steel or precast concrete deck, on the other hand, makes it more difficult, and often more frequent. The original economies offered by steel and other prefabricated units are canceled out by much higher roofing costs over a period of time. This chain of events if often the result of using an insulation material with highly praised thermal efficiencies but with other qualities that render it less than satisfactory for roofing work. Insulation materials that individually are fine products when used within their capabilities are not generally suitable for use in a built-up roof system that is subjected to extremes in temperature and ever-present moisture in vapor and liquid form. They are apt to endanger the system rather than enhance it.

Examples

The following examples show mistakes that were made in the structural design and roofing system design at great cost to the building owners.

Reroofing Procedures

Example 1
Office building. Toronto, Ontario.

178 squares of 4-ply pitch and gravel over 2-in. cork insulation. Asphalt-coated paper vapor barrier on a precast concrete deck. Roof applied November 1954. Leaks and blistering reported over the whole area in April 1966. Photographs and investigation of seven cut-outs 12 in. $\times$ 12 in. showed loss of gravel, bare felt, buckles, blisters, and cracks over most of the roof. The tests showed there were full quantities of materials in the roof membrane, except where wind erosion had removed the gravel, and a reduction in weight of pitch due to the usual evaporation of coal-tar constituents. The owner was advised to replace the entire system and change the parapet flashing design, where copper was flashed into the copings, and caulked with lead wool. This detail always leaked. The owner, however, decided to have the gravel removed and the surface resaturated (sic) and regraveled. The cost was $5,000 or $28 per square. Late in 1967 the roof was inspected and found to be covered with soft sticky coal tar and a great deal of loose gravel. In September 1971 the roof was replaced at a cost of $30,000 ($168 per square). Comment — $5,000 was wasted on coating. The asphalt-coated paper on a precast concrete deck could not possibly keep moisture vapor out of the system, which was principally pitch.

Example 2
Two identical building materials warehouses, one in Alberta and one in Ontario.

Smooth-surfaced asphalt felt roofs on sloping plywood decks. Roof system — dry sheet, two nailed felt plies and three mopped plies. Membrane moderately buckled in both directions due to expansion of nailed felts. Undoubtedly due to interior moisture during construction. Very little danger of leakage or increased buckling. The roof should have been mopped directly to the plywood deck. No repairs were suggested to the Alberta building. Information on the Ontario building came later after it was too late. The same roof coating company recommended resaturating and coating each building at a cost of $25,000 each. The Ontario building was done before the coating company could be stopped. Nothing was done to the Alberta building so $25,000 was saved. The manager of the Ontario building did not know he had been duped.

Example 3
Storage warehouse connected to paper mill machine room.

Shiplap deck — level. Roof system — dry paper, two plies of nailed asphalt felt and two plies mopped, plus a flood coat and gravel. Roof not deteriorated in any way but the ceiling dripped water. Cut-outs showed the unsaturated paper was wet but not the dry nailed felts. Moist air from the paper machine room was believed to be condensing on the underside of the roof. A roofing contractor had submitted a bid for $14,000 to install a new roof. A roof coating company suggested coating and regraveling for $18,000. Nothing was done to the roof. The space was properly ventilated and the dripping stopped.

Example 4
Plywood mill.

Two barrel roofs connected with a valley — total area, 1740 squares. Roofed in 1967. Total cost for barrel roofs and valley — $39,000. Roof specification — (for barrels) red resin paper on plywood deck, two plies of mopped asphalt asbestos felt and one ply of No. 80 white asbestos mineral-surfaced roofing—two mops of asphalt. (This is not a particularly good specification for the building.) In 1971 the roof was inspected because of buckles in the mineral-surfacing roofing parallel to the laps and near the edge. The roof was not leaking and never had. The buckles were found to be full of water even though the roof was steeply sloped. They were drained by opening them at the eaves. Water was believed to have entered the system from the interior due to construction-phase moisture not being vented off. This was confirmed after a discussion with the supervising engineer on the construction sequences. The buckles were also partly due to an extremely wet manufacturing operation. Ventilation of the building was inadequate because of high interior and exterior humidity levels. No roof repairs were suggested except draining the buckles.

Repairs suggested by roof coating company 1:

A. Repairs to blisters and aluminum coating. Ten-year guarantee. $27,254.
B. Apply three plies of asbestos felt and aluminum coating. $75,000.

Repairs suggested by roof coating company 2:

A. Cut and patch blisters and touch up with aluminum paint. Estimated (not firm) $13,200.
B. Same as A and add one ply of #30 saturated felt, nailed, and SIS roofing (17-in. slate or 16-in. selvage edge roofing) laid in hot asphalt. Estimated (not firm) $49,000.

None of the above repair suggestions would have solved the buckling problem and none would have corrected the basic error in the original roofing specification.

Example 5

Three identical apartment buildings built on the same site.
No.1 completed 1966. No.2 completed 1967. No.3 completed 1968. Reinforced concrete, 18 stories, plus four penthouses on the 19th floor. Penthouse roofs, 81 squares, level.

Roof system—No.1 (penthouse roof 1966).
Concrete deck, 2-in. expanded polystyrene, 3-ply felt and asphalt with gravel surfacing. No vapor barrier. Cost in 1966: $5,000
Failed in 1970 and reroofed with 4-ply asphalt felt and gravel over expanded polystyrene. No vapor barrier. Cost not known.
Failed in 1976 and reroofed with Uniroyal #6125 modified asphalt coating applied with squeegee and covered with one ply of No. 15 asphalt felt. Covered with 2-in. extruded polystyrene and gravel ballast. Cost: $28,000

Roof system—No.2 (penthouse roof 1967).
Concrete deck, 2-in. expanded polystyrene, 4-ply asphalt felt and gravel. No vapor barrier.
Failed 1982. Reroofed same as third roof on No.1. Cost: $58,000

Reroofing Procedures

Roof system — No. 3 (penthouse roof 1968).
Concrete deck, 1-in. extruded polystyrene and 1-in. polyurethane, 4-ply asphalt felt and gravel. No vapor barrier.[1]
Failed 1981. Reroofed same as No.1 and No.2. Cost: $39,000

Summary — Seven roofs between 1966 and 1982 (16 years) on three buildings. Total cost, including original roofs, approximately $155,000, but not including extensive and expensive repairs on all three buildings during that time, interior decorating repairs, damage to tenant's property, and loss of rent because penthouses were not habitable during part of those 16 years. Cost per square of roof to date: approximately $1,914.

Example 6

Shoe factory — Southern Ontario, 1966.

Concrete block walls, level steel roof deck on open web joists, fiberboard insulation and 4-ply asphalt felt roof with gravel surface. No vapor barrier. Building constructed and roofed in late fall. Interior ground excavated and concrete slab poured after roof was laid. Interior heated but not well ventilated. Moist air from curing slab passed through the steel deck and into the wood fiber insulation, which collapsed into the flutes in the steel deck. The roof followed, causing widespread leaks from a ponded roof. The building was started in September and shoes were being manufactured in January. A court of law ruled in favor of the owner, who sued the general contractor, the roofing contractor, and the roofing manufacturer for $25,000. The general contractor disappeared, the roofer could not be held responsible, and the manufacturer settled out of court for $10,000. The general contractor had supplied the entire building on a turn-key operation, but the manufacturer had supplied a 20-year bond on the roof before the leaks started. Before the roofer could be paid by the general contractor, the manufacturer had to produce the bond. The lack of air/vapor barrier hastened the destruction of the insulation but if one had been installed it probably would not have prevented the problem for many years. It might, however, have delayed the hasty departure of the general contractor.

Example 7

A similar case but involving a much larger building occurred in Ontario in 1964–65.

A 3-ply pitch and gravel roof of 1,960 squares was laid over an asphalt-coated base sheet on 1 1/2 in. of insulation and no air/vapor barrier. The deck was steel. Work proceeded during the winter and the roof leaked during and after construction was completed. The roof eventually failed completely and was replaced in 1970 at a cost of $190,843. The original cost of the roof is not known but was included in the $1,611,000 building cost.

The roofing contractor warned the general or prime contractor and the design engineer as follows but was ignored:

> There is no vapor barrier under this roof and there is therefore considerable danger of moisture from inside the building passing through

---

[1] The polyurethane was saturated with water.

the joints of the roof deck and getting into the roof insulation. This could give rise to a very serious condition later on. Giffels Associates to be asked to consider this problem.

A further quote from the summation follows:

> His Lordship reviewed the evidence and found as a fact that the failure of the roof was caused by the conduct of the prime contractor in providing inadequate ventilation while drying out the concrete floor, thus causing excessive moisture to condense on the underside of the roof and ultimately to seep into the roof insulation. He then continued:
> The responsibility for the substantial failure of this roof within five years of its completion is that of Eastern and Giffels. Eastern took a calculated risk by proceeding with thawing the frozen ground and pouring concrete floors in winter conditions using artificial heat and furthermore prevented this heat from being dissipated by keeping the building enclosed at all times. High humidity and condensation resulting in moisture destroyed the effectiveness of the insulation as previously described. Notwithstanding the complaints and warnings from the roofer, no steps were taken to remedy these conditions. Eastern continued construction regardless of the conditions that were being created in order to meet the completion date.

The roofing specification supplied by the owner's engineer was a strange mixture of asphalt and pitch and an asphalt-coated base sheet. The origin of the specification is not known for sure but is suspected. It is evident the engineer had no knowledge of such matters, and it is surprising the roofer did not object to it. The insulation is described as being in two parts, the first 1 in. thick and the second, $1/2$-in. fiberboard. The material in the first layer is not identified nor is its attachment to the steel deck.

When the smoke cleared away, the judgment amounted to $107,121. The roofing contractor won the case. The owner was claiming $218,388, plus interest.

There are times when the construction schedule interferes with good building practice and while this might have been the case in this example, it is doubtful. In the building in Figures 10.4 and 10.5 the critical path scheduling came very close to causing chaos during the final stages of roofing when two cranes were taken down a few days earlier than was convenient for the roofer. The general contractor delayed the roofer by leaving sections of the deck open for installation of equipment. It was also discovered by accident that the crane workers drilled eight holes through the insulation, roof membrane, and deck without notifying anyone. They were installing two temporary devices to assist in the crane removal. In a conventional system this would have destroyed the integrity of the membrane and filled the system up with water.

Urethane Foam

More correctly known as polyurethane, urethane foam *is an insulating material, not a roofing material.* It is made from two components: a polyisocyanate and a

Rerooting Procedures

polyisocyanate and a polyhydroxyl. The components also contain catalysts, a surface-active agent, and a blowing agent, which is normally a fluorocarbon refrigerant. The refrigerant gases have large molecules that do not transfer heat readily and are easier to hold in the plastic material. Normally, refrigerants R-11 and R-12 are used. Mixing of components can be done in the factory or sprayed on site. Factory-made foam has greater thermal resistance than sprayed foam because of the shape of the cells and has more resistance to aging. Sprayed foam is usually 2 to 3 lb per cubic foot in density. It is used **as insulation** on roofs too difficult for sheet materials or traditional roofing and on old roofs where the thermal resistance of the roof system must be improved. It must be covered with a coating that is waterproof, durable, and resistant to ultraviolet radiation. Materials used for this purpose are silicone, acrylic elastomers, butyl elastomers, hypalon, neoprenes, vinyls, and polyurethanes. They are seldom more than 20 or 30 mils thick.[2] Aluminum coatings are used but they are much thinner. The foam has good chemical resistance but has severe shortcomings in many other respects. These include:

1. High flame spread relative to other building materials. Large amounts of smoke and dangerous fumes can be generated. Increasing fire resistance by adding phosphorous, chlorine, or bromine also increases cost, friability, and brittleness, and in some cases, reduced chemical resistance. ASTM standard D-1692-59T and 67T.
2. High resistance to water vapor transfer, but it must be covered with coatings and protected by vapor barriers against water absorption and dimensional instability.
3. Cost on the basis of cents per unit of thermal resistance is five to seven times that of glass fiber insulation and two or three times that of polystyrene. The application cost is low.
4. Not resistant to ultraviolet radiation and dimensionally unstable. When relative humidity is more than 90% and temperature exceeds 120° to 150°F, the linear expansion is 5% to 20%, and by volume 15% to 60%. When dry the material is more or less dimensionally stable, with under 5% change in each direction and under 15% by volume — temperatures up to 220°F. **Polyurethane is best used where it can be kept cool and in a dry environment. It should be protected from extremes in temperature and especially from wetting. This is not roofing country.** White roof surfaces can reach 130°F.
5. Spray-on urethane has spherical cells, elongated in the direction of the rise. This reduces the thermal resistance and vapor resistance and increases the rate of aging.
6. Polyurethane will disintegrate after only a few freeze-thaw cycles.
7. At temperatures below +50°F, the refrigerant gas in the cells will condense causing some small shrinkage and reduced resistance to heat flow. The thermal resistance of aged material is somewhere between 5 and 6 depending on temperature. The lower the temperature the less efficient it is.

---

[2]The average thickness of most single-ply sheet membranes is 50 mils.

8. Patching damaged or defective coatings and the polyurethane foam may be difficult due to a dirty surface and crumbled foam. If portable repair kits are inadequate, all the original application equipment must be returned to the building. It has been reported that silicone coatings can turn black due to an electrostatic property that attracts dust. This increases the surface temperature. The difference between a pure white surface and a black one is about 60°F.

9. Long-wave radiation from reflective roof surfaces makes it extremely uncomfortable to walk or work on the roof. Special footware, tinted goggles, and white clothing are suggested. Short-wave radiation from the sun does not pass through heavily insulated roof systems. Up to 80% is reflected back from a white surface as long as the surface stays bright and clean.

Urethane Foam

For more complete information on polyurethane as a thermal insulation, refer to C.J. Shirtliffe, Building Research Note No. 124, November 1978. National Research Council of Canada. ISSN 0701-5232, and the Urethane Institute, The Society of the Plastic Industry, Inc., 355 Lexington Ave., New York, N.Y. 10017.

17

Single-Ply Membranes

Introduction

Now that we are discovering how little we know about conventional built-up roofing, we are faced with hundreds of new materials, for which until recently, there were no standards. Manufacturers provided a long list of physical properties, determined by standard test methods, but often having little or no relation to the performance of the material for roofing. Chemical formulations and recommended minimum thicknesses for some materials also appear to change overnight, and you cannot be sure that the material you see at some point in time is the same material you examined two weeks earlier. It is hard to believe that designers and roofers would be willing to take a chance on such materials, in the absence of standards and specifications, but such has been the case for 15 or 20 years. During that time, materials have come and gone leaving a trial of distress and lawsuits behind them, and still they come. It is strongly recommended that any designers or roofers who use these materials in the absence of a proper material standard, test method for evaluation, and specification for use by an independent standards writing agency, be very careful with their clients' money. Certainly not all new materials or systems are bad, but enough of them are unsuitable for the environmental conditions in which they have to serve to require careful consideration. And the record of achievement for even the better ones is not entirely without blemish.

Single-Ply Membranes

The current interest in single-ply elastomeric and modified bituminous membranes is partly due to (1) the failure rate of conventional roofing systems and the inconvenience of hot application, and (2) the glamour aspects of roofing materials and practices in other countries, and the readiness of importers and local manufacturers to exploit something that is new. Extraordinary performance reported in Europe and Japan is not easy to confirm and may not be repeated in North America on different construction, with completely different labor practices and different expectations.

Manufacturers

The last count by the NRCA indicated there were approximately 55 U.S. manufacturers or agents marketing sheet materials. Some of these are listed below. Naturally they are all confident in their advertising that their particular product is the solution to all roofing problems. Enthusiasm may sell products but it does not begin to reach the root of the problems — and there are many. Table 17.1 lists manufacturers of six product groups and 31 individual sheet roofing items, numbered and identified by name in Table 17.2. Tables 17.3A and 17.3B show 12 product properties and details of application. This information is supplied by the manufacturers but may have been revised. In each column the variety of detail between products makes it difficult to decide which is the best for any particular purpose. The information shown is only about one-half of what is in the original source.[1]

Some items are worth noting.

Except for the modified bitumen sheets that are up to 4.0 mm thick, the rest are only slightly more than 1.0 mm thick. There are several forms of lap cementing and four methods of securing or anchoring the sheet to the substrate. There are few exceptions to approval of all the presently popular roof insulations to be used as a base (see Table 17.4). There is no mention of any other base like concrete, wood, or plywood. There is also approval of a dead-level roof deck except in four cases, all modified bitumens.

Product 13 — 1% slope.
Product 15 — 2% slope.
Product 18 — 4% slope.
Product 19 — 4% slope.

Three weaknesses are evident:

A dead-level roof would pond with water and promote undesirable roof traffic.

A very thin sheet over soft low-density insulation would permit only the lightest form of roof traffic. Puncturing of the sheet is probable.

Ponded water over a narrow cemented lap is not encouraging. It is the antithesis of wide laps in hot roofing.

---

[1]*RSI 1982 Handbook of Single-Ply Roofing Systems.* Reproduced with permission of Roofing/Siding/Insulation, 757 Third Avenue, New York, N.Y. 10017.

Table 17.1

Generic Material	Manufacturer	Product number
CPE Chlorinated Polyethylene	Cooley Roofing Systems, Inc.	1
EPDM Ethylene Propylene Dienne Monomer	Carlisle Tire & Rubber Co.	2
	Celotex Corp.	3
	Firestone Industrial Products Co.	4
	Gates Engineering Co. Inc.	5
	GTR Building Products Co.	6
	Goodyear Tire & Rubber Co.	7
	Manville Building Materials Corp.	8
	SY Energy Methods, Inc.	9
Hypalon Chlorosulfonated Polyethylene or CSPE	J.P. Stevens & Co. Inc.	10
	Tremco, Inc.	11
	Uniroof/Central States Assoc. Corp.	12
Modified Bitumen	Imper Italia S.P.A	13
	U.S. Intec	14
	Koppers Co. Inc.	15
	Monroe, Inc.	16
	Rhoflex Div. of Teltex, Inc.	17
	Siplast, Inc.	18
	Tamko Asphalt Products, Inc.	19
Neoprene Polychloroprene	Gates Engineering Co. Inc.	20
PIB Polyisobutylene	AGR Co.	21
	Gates Engineering Co. Inc.	22
	Republic Powdered Metals, Inc.	23
	Tropical Industrial Coatings, Inc.	24
PVC Polyvinyl Chloride	Dynamit Nobel of America	25
	GTR Building Products Co.	26
	ipw Interplastic	27
	Rubber & Plastics Compound Co. Inc.	28
	Sarnafil (U.S.) Inc.	29
	United States Mineral Products Co. Weather Shield Division	30
	Flag-Wat Pro	31

The use of contact adhesives with an instantaneous grab is tricky in the best of circumstances. With wide flexible sheets that tend to buckle and ripple while being laid, the use of this type of adhesive is questionable and would require extremely careful and skilled workmanship.

The selection or suggestion of a suitable insulation substrate appears rather odd, e.g., complete acceptance of wood fiber and expanded polystyrene, both more or less rejected in roofing work, and is difficult to understand. Only 6 out of 31 suggest extruded polystyrene which is infinitely superior to expanded polystyrene, and which is the only acceptable insulation for use in the Protected Membrane

Single-Ply Membranes

Table 17.2

Product Number	Product Name
1.	Cool Top 40
2.	Sure-Seal Design B System
3.	Celo−1
4.	Rubber Gard
5.	Gacoflex System 1-B/E-2S
6.	Genflex ACR Roofing System
7.	Versigard Roofing System
8.	SPM Roofing System
9.	SMI Rubber Roof Energy Systems
10.	Hi-Tuff Roofing System
11.	Tremply
12.	Uniroof
13.	Pralon NT4
14.	Brai SP4
15.	KMM Aluminum Membrane
15.A	KMM Standard Membrane
16.	Rhinohide
17.	Rhoflex
18.	Parafor−50
19.	Awaplan
20.	Gacoflex System 11/N−35
21.	Alphagard Roofing System
22.	Gacoflex System 111/P−4S
23.	Geoflex P.I.B.
24.	Tropigard Roofing System
25.	Trocal Roofing Membrane
26.	Genseal ACR Roofing System
27.	Interoof
28.	Nervaply
29.	Sarnafil
30.	Flexhide
31.	Flagon

System (No. 4 in the roof system column). The inclusion of foamglas for product 2, which is only 1.1 mm thick, must be an error. The thin membrane would be cut through on the sharp corners of the foamglas in a very short time.

In nearly every case contaminants to avoid include petroleum products such as oil, grease, solvents, acids, and hydrocarbons such as asphalt and coal-tar pitch. These materials are present on nearly all level or near-level roofs on commercial and industrial buildings; therefore single-ply membranes on these structures should be carefully examined for suitability.

While external contaminants may damage single-ply membranes, the solvents in liquid-applied materials and the solvents in some adhesives for sheet materials may dissolve or damage foamed plastic insulations, particularly polystyrene. Also, the toxicity and flammability of solvents may pose a minimum risk to the worker on an open roof.

Insulation

Approval by manufacturers of single-ply membranes for new roofing, reroofing, and over various forms of thermal insulation is shown in Table 17.4.

Table 17.3A
Partial Product Specifications from Manufacturer

Product No.	Thickness Mils (mm)	wt/ft²	Roll Size ft in.	Tensile Strength p.s.i.	Elongation %	Min. slope
1.	40 (1.0 mm)	0.31	62 × 103.5	2500 ASTM D 852	27 ASTM D 751	level
2.	45 (1.1 mm)	0.28	50) 4.5 ft 100) through 125) 45 ft	Typical 1640 ASTM D412	500 ASTM D412	level
3.	45 (1.0 mm)	0.28	10 × 100 ft 20 × 100 ft	1400 min. ASTM D412	300 ASTM D412	level
4.	45 (1.1 mm)	0.28	40 × 200 max.	1650 ASTM D412	450 ASTM D412	level
5.	45 (1.1 mm)	0.26	60 × 100	1400 ASTM D412	300 ASTM D412	level
6.	60 (1.5 mm)	0.35	20 × 100	1450 ASTM D412	255 ASTM D412	level
7.	45 (1.1 mm)	0.27	4.5 – 24 × 100	1300 ASTM D412	300 ASTM D412	level
8.	45 (1.1 mm)	0.245	up to 39 × 125	1400 ASTM D412	350 ASTM D412	level
9.	55 (1.38 mm)	0.40	10,20,52 × 104 – 156	1400 ASTM D412	300 ASTM D412	level
10.	45	0.29	58 × 75	1400 ASTM D412 w.o. scrim (10)	400 min. ASTM D412 w.o. scrim	level
11.	40 (1.0 mm)	0.33	5 × 120	1458 ASTM D412	748 ASTM D412	level

Table 17.3A (cont'd)

Product No.	Thickness Mils (mm)	wt/ft²	Roll Size ft in.	Tensile Strength p.s.i.	Elongation %	Min. slope
12.	35 (5) (0.88 mm)	0.25	39 × 82	1000 min. ASTM 2707	400 ASTM 2707	level
13.	160 (4.0 mm)	0.78	3'–3" × 32'–10"	500 L (6) 390 T (UEA tc)	50 L&T (6) (UEA tc)	1% (UEA tc)
14.	150 (3.75 mm)	0.82	3'–3" × 32'–9"	400 Din 53857	50 Din 53857	level
15.	120 (3.0 mm)	0.92	43.3" × 33.1'	282.5 ASTM D412	226 ASTM D412	2%
15A	160 (4.0 mm)	1.2	43.3" × 33.1'	66.7 ASTM D412	408 ASTM D412	level
16.	160 (4.0 mm)	0.78	3'–3" × 33'–0"	500 L (6) 390 T	50 (6) L&T ASTM D 2523	level
17.	160 (4.0 mm)	0.88	3'–3" × 33'–0"	550 ASTM D412	57 ASTM D412	level
18.	150 min. (3.75 mm)	1.47	3.28 × 17.75 ft	170.3 min. CSTB 11-78	40 CSTB 11-78	1/2 in 12 4%
19.	160 (4.0 mm)	1.05	3.28 × 33.9 ft	130 ASTM D2523	50 ASTM D2523	1/2 in 12 4%
20.	60 (1.5 mm)	0.452	48 in. × 100 ft	1800 ASTM D412	350 ASTM D412	level
21.	100 (2.5 mm)	0.57	42 in. × 50 ft	550 ASTM D412 & D882	410 with backing	level

22.	100 (2.5 mm)	0.57	42 in. × 50 ft	550 ASTM D412 & D882	410 with backing ASTM D882	level
23.	100 (2.5 mm)	0.57	42 in. × 50 ft	550 ASTM D412 & D882	410 with backing ASTM D882	level
24.	100 (2.5 mm)	0.57	42 in. × 50 ft	550 ASTM D412 & D882	410 with backing ASTM D882	level
25.	48 (1.2 mm)	0.35	varies	>2400 ASTM D638	>250 ASTM D638	level
26.	45 (1.1 mm)	0.30	70 in. × 72 ft	2100 ASTM D412	300 ASTM D882	level
27.	60 (1.5 mm)	0.40	72.8 in. × 67.9 ft	2850 Din 53455	360 Din 53455	level
28.	48 (1.2 mm)	0.38	54 in. × 80 ft	2390 ASTM D882	325 ASTM D882	level
29.	48 (1.2 mm)	0.33	6.5 ft × 65 ft	1400 ASTM D882	220 ASTM D882	level
30.	45 (1.1 mm)	0.30	82 in. × 75 ft	2330 ASTM D882	400 ASTM D882	level
31.	48 (1.2 mm)	0.33	6.5 ft × 65 ft	2600 ASTM D882	350 ASTM D882	level

Table 17.3B
Partial Product Specifications from Manufacturer

Product No.	Lap jointing (9)	Low temperature flexibility °F	Workable temperature range °F	UV resistance	Type of roof system (2) 1	2	3	4
1.	Hot air welding	−40 Mandrel test	20 to 120	5000 hrs. (1)	—	X	—	—
2.	Cement and sealant	−75 ASTM D746	−69 to 180	2000 hrs. ASTM D2565	X	—	—	—
3.	Contact adhesive	−62 ASTM D746	−62 to 300	10 years ASTM D1149	X	X	X	X
4.	Synthetic polymer	−75 ASTM D746	−40 to 180	166 hrs. 50% extension. 100F 2000 hrs. ASTM D1149	X	X	X	X
5.	Contact adhesive	158 (sic) (3) ASTM D746	50 to 90	5,000 hrs. ASTM D1149	X	—	—	—
6.	3" wide seam. Contact adhesive (7)	−67 ASTM D746	40 to 80	no cracks ASTM D1149	X	X	X	X
7.	Contact adhesive	−67 ASTM 2137	40 to 90 ideal	excellent ASTM D1149	X	X	X	X
8.	Contact cement	−75 ASTM D746	40 to 80	excellent ASTM D750	X	—	—	—
9.	Synthetic rubber adhesive. Plate bonding (7)	−70 ASTM D2137	32 to 120	no cracking ASTM D1149 (4)	X	X	X	X
10.	Hot air solvent welding	−40 ASTM D2136	10 to 110	Excellent 15 yr. Panama weathering	X	X	—	X

#	Description	Value	Temp range	Notes					
11.	Latex rubber contact cement. Solvent weld	60 ASTM D746	20 to 110	No cracks 18 yr. weatherometer. ASTM E42	—	—	—	X	—
12.	Uniroof AD 432 neoprene latex	−40	10 to 120	Excellent ASTM G23	—	—	—	X	—
13.	Torch welded (7)	14 (UEA tc)	20 to 120	Excellent 500 hrs. QUV tester ANSI-ASTM G53-77	X	X	—	X	—
14.	Torch welded over expanded polystyrene & urethane (7)	−4 (AB test)	0 to 120	Membrane protection recommended e.g., alum.	X	X	—	—	X
15.	Asphalt solvent (7)	22 ASTM D228	10 to 120	covered with alum. foil	—	—	—	X	—
15A	Heat fusion on wood fiber & polystyrene (7)	14 ASTM D228	10 to 120	Protected by ballast	X	—	—	—	X
16.	Torch welded	180° bend @ −14 F	35 to 120	No change after 500 hrs. ASTM G53-77	—	X	—	X	—
17.	Torch welded (7)	0° ASTM D746	0 to 130 air temp.	No deterioration. 672 hrs. accelerated thermo hydro shock testing	—	X	—	X	—
18.	Hot asphalt or torched PA-311 adhesive	5 CSTB 11-78	15 min. to 120	opaque granular covering	—	X	—	X	—
19.	Hot asphalt ASTM D312 Type 3	0 ASTM D2136	50 to 120	opaque granular covering (8)	—	—	—	X	—

Table 17.3B (cont'd)

Product No.	Lap jointing (9)	Low temperature flexibility °F	Workable temperature range °F	UV resistance	Type of roof system (2)			
					1	2	3	4
20.	Mechanically fastened and contact adhesive	−42 ASTM D746	50 to 90 Material temperature	Excellent 21 yrs. Exp. in Panama	—	X	X	—
21.	Approx. 50% bonding with asphalt or contact adhesive	−40 ASTM D2137	20 to 100	meets Din 4102	X	X	—	X
22.	Spot mopped hot asphalt. Prefab sealing strip	Ditto	Ditto	Ditto	—	X	—	—
23.	Ditto	Ditto	Ditto	Ditto	X	X	—	X
24.	Ditto	−58 ASTM D746	Ditto	Ditto	X	X	X	X
25.	Mechanically fastened (7)	−60 ASTM D2137	−15 to 120	10,000 hrs. 1,200 kw/hr	X	X	X	—
26.	Mechanically fastened (7) (9)	−40 Din 53372	40 to 80	600,000 emmaqua	X	X	—	X
27.	PVC coated strips (7) Polypropylene non-woven	−30 ASTM D746	5 to 120	2,000 hrs. xenon test type 250 (Din 53387)	—	X	—	—
28.	Solvent welded to batten strips THF or hot air (7)		30 to 120	3,000 hrs. ASTM E42	X	X	—	—

29.	(2) Polyester reinforced membrane. (3) Sarnacol water & solvent based (7)	−40 ASTM D746	0 to 120	5,000 hrs. xenon exp.	X	X	X
30.	Polyester, aluminum foil, glass, microfoam. (7)	−40 ASTM D746	Above 40	Elongation after three million Langleys min. 75% of original.	X	—	X
31.	(1) Pea gravel 6 psf. Flagon C (2) Mech. fastened. Flagon MF (3) Hot asphalt, cold emulsion-Flagon SF	−40 ASTM D746	20 to 120	Excellent Xenon test 150	X	X	—

[1] Twin carbon arc weatherometer (federal standard 195–method 5804)
[2] Roof systems: (1) loose laid ballasted
(2) Partially attached
(3) Fully adhered
(4) Protected membrane roofing assembly
[3] Obviously an error
[4] 30 days 50 pphm–100°F. 20% elongation
[5] Fiber felt backing
[6] L - longitudinal
T - transverse
[7] Asphalt-saturated base sheet or other separator sheet
[8] Asphalt emulsion coating required for UL Class A
[9] Lap jointing and adhesive applies to type of roof system No. 2 only, unless otherwise indicated
[10] w.o. scrim–without scrim

Table 17.4

	1	2	3	4	5	6	7	8	9	10	11	12	13	14	15	15A	16	17	18	19	20	21	22	23	24	25	26	27	28	29	30	31
New roofing	X	X	X	X	X	X	X	X	X	X	X	X	X	X	X	X	X	X	X	X	X	X	X	X	X	X	X	X	X	X	X	X
Reroofing	X	X	X	X	X	X	X	X	X	X	X	X	X	X	X	X	X	X	X	X	X	X	X	X	X	X	X	X	X	X	X	X
Perlite	X	X	X	X	X	X	X	X	X	X	X	0	X	X	X	X	X	X	X	X	X	X	X	X	X	X	X	X	X	X	X	X
Urethane	X	X	X	X	X	X	X	X	X	X	X	X	X	X	X	X	0	X	0	X	X	X	X	X	X	X	X	X	X	X	X	X
Wood fiber	X	X	X	X	X	X	X	X	X	X	X	X	X	X	X	X	X	X	X	X	X	X	X	X	X	X	X	X	X	X	X	X
Fiberglass	X	X	X	X	X	0	X	X	X	X	X	X	X	X	X	X	X	X	X	X	X	X	X	X	X	X	0	X	X	X	X	0
Expanded polystyrene	X	X	X	X	X	X	X	X	X	X	X	0	X	0	X	X	0	0	X	X	X	X	X	X	X	X	X	X	X	X	X	X
Extruded polystyrene	X	0	0	0	0	0	0	0	0	0	0	0	0	0	X	0	0	0	0	0	X	X	0	0	0	X	X	X	0	0	0	0
Felt-faced composites	X	X	X	X	X	0	X	X	X	X	X	X	X	X	X	X	X	0	X	X	X	X	X	X	X	X	X	X	X	0	X	X
Cellular glass	0	X	0	0	0	0	0	0	0	0	0	0	0	0	0	0	0	0	0	0	0	0	0	0	0	0	0	0	0	0	0	0

X - approved   0 - not approved

Contaminants

Manufacturer's suggestions for contaminants to be avoided are shown in Table 17.5.

Table 17.5

Product number	
1.	Strong oxiders, aromatic hydrocarbons
2.	Most oils – check manufacturer
3.	Some oils – check manufacturer
4.	Aliphatic and aromatic hydrocarbons, lubrication and compressor oils
5.	Oil base or plastic roof cement, waste products, steam
6.	Acids, animal fats, fresh asphalt
7.	Petroleum-based products
8.	Solvents, oils, grease, fresh coal tar, roof cement
9.	Toulene oil, grease – consult manufacturer
10.	Aromatic hydrocarbons, oxygenated solvents
11.	Chlorinated and aromatic solvents
12.	Hydrocarbons
13.	Organic compounds, e.g., oil solvents
14.	Concentrated acids and hydrocarbons
15.	Petroleum solvents, some chemicals
15A.	Petroleum solvents, some chemicals
16.	Petroleum solvents
17.	Petroleum solvents – check with Teltex
18.	Coal-tar derivatives
19.	Coal tar, aromatic solvents
20.	Aromatic solvents, strong oxidizing chemicals
21.	Tar, solvents, organic oil and fats
22.	Ditto
23.	Ditto
24.	Ditto
25.	Check manufacturer's chemical resistance list
26.	Petroleum-based products
27.	Bituminous coal tar
28.	Fresh coal tar
29.	Bitumen, oils, animal fats
30.	Coal tar, asphalt, oil, solvent
31.	Asphalt, coal-tar pitch

Single-Ply Membranes

Canadian General Standards Board
CGSB Standards for New Roofing Materials

Table 17.6
Canadian General Standards Board
CGSB Standards for New Roofing Materials* 1982

37-GP-50M	Hot-applied rubberized asphalt (issued March 1978)
37-GP-51	Application of hot-applied rubberized asphalt (issued December 1979)
37-GP-52M	Prefabricated elastomeric sheeting (approved in committee, prepublication issue August 1983)
37-GP-53	Application of prefabricated elastomeric sheeting (1st draft) May 1980
37-GP-54M	Sheet-applied flexible PVC (issued January 1979)
37-GP-55M	Application of flexible PVC (issued September 1979)
37-GP-56M	Prefabricated, reinforced, modified, bituminous membrane (issued July 1980)
37-GP-57 (Presently as an Appendix to 37-GP-56M)	A guideline for the application of prefabricated, reinforced, modified, bituminous membrane (first draft) November 1980
37-GP-58M	Cold-applied, liquid elastomeric membrane (non-exposed) (fourth draft) November 1980
37-GP-59M	Cold-applied, liquid elastomeric membrane (exposed – non-traffic-bearing areas) (second draft) July 1980
37-GP-60	Cold-applied, liquid elastomeric membranes (exposed – traffic-bearing areas) (early stage)
37-GP-61	Application of cold-applied, liquid elastomeric membranes (early stage)

*Canadian Government Publishing Centre
Supply and Services Canada
Ottawa, Canada
K1A 0S9   Telex 053-4296   Tel. (819) 997-2560

References

For more complete information on elastomeric roofing, the following sources may be consulted:

1. Rossiter, Walter, Jr., and Mathey, Robert. *Special Report*, R/S/I Aug., Sept., Oct., Nov. 1979. National Bureau of Standards.
2. The Roofing Industry Educational Institute. *Non-Conventional Roofing Systems.* 6851 S Holly Circle, Suite 250, Englewood, Colo. 80112.
3. Single-Ply Roofing Institute, 1800 Pickwick Ave., Glenview, Ill. 60025.
4. Midwest Roofing Contractors Association (MRCA), 1000 Power & Light Building, Kansas City, Mo. 64105.
5. Canadian General Standards Board, CGBS Standards for New Roofing Materials, Canadian General Standards Board (sales), c/o Department of Supply & Services, Place du Portage, Phase III—4 BI, 11 Laurier Street, Hull, PQ Canada K1A 0S5.
6. *1983 Handbook of Single-Ply Roofing Systems, Roofing/Siding/Insulation*, Harcourt Brace Jovanovitch Publications, 757 Third Ave, New York, N.Y. 10017.
7. National Bureau of Standards. National Engineering Laboratory, Gaithersburg, Md.
8. A.S.T.M. Committee D-08. *Bituminous and Polymeric Materials for Roofing, Waterproofing and Industrial Uses.*

18

Cold Built-up Roofing

History

The application of roofing using cold adhesives started in the infancy of asphalt roll roofing early in the twentieth century. The coated roofing was applied to steeply sloped roof decks and lapped 2 in. A cold-solvent-type (cut-back) asphalt lap cement, often packed in the roll with nails, was applied with a brush or stick to the lap and the nails driven through the overlapping sheet. Both smooth-surfaced and mineral-surfaced roll roofing were applied in this manner. The cement generally did not contain any stabilizing fiber and often ran down the roof. The principal disadvantage to this application was the withdrawal of the nails from the wood deck. Even though the nails were only $7/8$ in. (2.24 cm) long, they would pop out of green lumber.

    A better method was developed later in which the roofing was lapped 4 in. (10 cm), blind nailed, and cemented. The cemented lap helped prevent nail pop, and even when it did occur the roof did not leak at the nails. It should be mentioned that the liquid lap cement was changed to an asbestos fibrated asphalt gum, which was a better adhesive and would not run down the roof when heated by the sun.

    The next step was the development of a 19-in. selvage-edge type of roofing described in chapter 6. This material was made in different ways by different manufacturers. Essentially it had a 19-in. selvage edge and 17 in. of mineral

Cold Built-up Roofing

surfacing on a 36-in.-wide sheet. It was known as a split sheet, N.I.S., or S.I.S. (N.I.S. is an abbreviation for nineteen-inch selvage, and S.I.S. for seventeen-inch slate). The selvage edge and the back of the sheet might or might not be asphalt coated. If not, the material weighed 55 lb per roll or 110 lb per square, laid. If coated, the material would weigh at least 60 lb per roll. The extra 5 lb is not much to coat 165 ft² of material.

$$\frac{(36 + 19)}{12} 36 = 164.8$$

Uncoated selvage-edge roofing is recommended by manufacturers for hot application and coated material for cold application. The reason for this is that hot asphalt will combine uncoated felt, but uncoated felt will absorb the solvents in cold adhesives and reduce the adhesive properties. Because of this, many uncoated selvage-edge roofing jobs either leaked or failed owing to blowoffs or severe shrinkage in the cross-machine direction.

The next step on the road to cold application of coated roofing was the use of hot asphalt to combine two plies of coated roofing felt weighing approximately 70 lb per roll. The top surfacing might be left off entirely, or it might be a cold asphalt emulsion, a reflecting coating, or a hot-pour coat and gravel. Experience shows that this roofing system has not always been successful, particularly when laid over insulation.

Selection by Geographic Location

It is essential that coated roofing in cold-applied systems be warm at the time of application. If it is not warm, it will not lie flat and air pockets and wavy edges will result. Unless the air temperature is above about 60°F (15.56°C), the roofing must be stored in heated space and the rolls heated all the way through, not just the outside convolutions. Further precautions must be taken as described later in this chapter.

Product Description

Coated roofing used in cold application (also called *cold process*) is a double-coated sheet weighing between 40 and 60 lb per 100 ft². It is generally made with organic felt saturated with asphalt, coated both sides with filled (stabilized) asphalt and surfaced with slate flour or mica to prevent sticking in the rolls. There is little difference, if any, between this and any good-quality smooth roll roofing.

The cold cement is asphalt in the type II range, cut back with approximately 30% petroleum solvent to a brushing or spraying consistency. There may be a small amount of fiber (7% to 10%) to make it more stable on inclined surfaces. Too much fiber makes it hard to pump or spray.

When a cold roof is surfaced for additional weather protection, it may be sprayed with cut-back asphalt or asphalt emulsion, containing bentonite clay, and covered with roofing granules, the same as an asphalt shingle.

Cut-back asphalt cements and gums cure slowly by evaporation of the solvent. Depending on the porosity of the covering material, the temperature, and the solvent itself, it may require about 30 days to set up. This slow cure is not as desirable as a hot asphalt, which achieves maximum adhesion as soon as it cools, which is only a few minutes.

Asphalt emulsions cure by evaporation of water, which is about 40% of the total weight. They are, therefore, used only as coatings, never as cements, at least not in roofing. Emulsions cure slowly in humid atmospheres if the surface cures first, leaving water below the surface. Because of the water content, asphalt emulsions cannot be applied if there is any possibility of freezing or rain. If emulsions are frozen in storage, they cannot be used. If frozen after application, the material will not cure properly. If the surface cures rapidly first, you cannot walk on it for days. If it rains before it is properly cured, it will wash off the roof and down the drains. Outside of these disadvantages, asphalt emulsions are an excellent surfacing material for a smooth-surfaced asphalt roof.

The manufacturer or suppliers of cold-application materials must be aware of the compatibility between the coating asphalt on the roofing felt, the asphalt in the cement, and the asphalt in the emulsion surface coating. They may all come from different sources.

Roof Decks and Substrates

Cold-application roofs are not recommended for use directly on insulation, nonnailable decks, or inclines below 1 in. per foot. This means that their use is limited to the following decks:

1. Boards
2. Plywood on joists
3. Plywood on T&G decking
4. Nailable precast decks
5. Nailable poured decks
6. Wood fiber and cement
7. Over old, smooth-surfaced roofs (see below)

Because of the low cured volume in a brushed or sprayed coat of cut-back cement, the roof deck must be prepared carefully to provide a perfectly smooth surface to avoid air pockets between the first ply of roofing and the deck and between subsequent plies. There is nothing wrong with thin cement coatings as long as there are no gaps and there is intimate contact between the materials being bonded together. Cold-application roofs are sometimes laid over old graveled roofs after the gravel is spudded or scraped off, or over old smooth-surfaced asphalt roofs. Such roofs must be dry and free of all dust and loose material and contain no moisture in the system. Attachment may be made by nailing, cementing, or both, depending on circumstances. There is always the risk of blistering between the old roof and the new.

A cold-application system may be adopted to cover an old roof of less than 1 in. per foot incline simply as a matter of expediency or because a hot kettle or tank truck is not permitted in the area. It should be remembered, however, that heat is sometimes required to make the cement workable, and that the solvent is flammable

Cold Built-up Roofing

and explosive—more so than straight roofing asphalt. Further heating may be required to soften the roofing, and propane- or kerosene-heated rollers are sometimes used to soften both roofing and cement after being laid in order to achieve a satisfactory bond. One might ask, "Is cold-application roofing cold after all?"

Roof System Specification and Application

Cold-applied roofs are laid two ply (19-in. lap) or three ply ($24^{2}/_{3}$-in. lap) with approximately 1.5 U.S. gallons of cold asphalt adhesive per square between each ply. Surfacing can vary as described earlier in this chapter. After the solvent is evaporated out, there is approximately 25 lb of asphalt and fiber adhesive left in three layers of 100 ft^2. With only 8 lb per square in each layer to fill all voids, it is imperative that the deck be smooth and the coated felts carefully rolled into the cement. In a hot-applied roof, the net weight of asphalt in each mopping is approximately 20 lb per square, or two and a half times as much. The cement is most likely to be sprayed rather than applied by brush, roller coater, or squeegee.

An emulsion coating alone will vary from 2 to 4 gal per square and should be applied in one coat, because it may not be possible to walk back over it for several days until it is completely cured.

If mineral granules follow the sprayed coating immediately, limited traffic is possible. The weight of granules per square is between 18 and 25 lb. When granules are used, a cut-back or emulsion coating will not exceed 2 gal per square.

On inclines above 1 in. per foot, coated cold-application felts should be nailed through steel nailing discs at the upper edge of each ply to prevent sliding, and on all roofs where the felt is not cemented to the deck. The number of nails and size of discs will depend on the conditions that prevail on individual roofs. Nailing through insulation is not recommended.

Uncoated roofing felt is rolled directly from the roll into hot asphalt, but a roofing felt coated both sides with filled coating should be rolled out and allowed to lie flat for several hours to remove the curvature of the roll and to allow the material to adjust in dimension to ambient conditions. Some authorities suggest each roll be cut into two or three lengths. If the roll has been in warm storage and is then laid out on a cold roof deck, the warming process is wasted. Furthermore, the labor involved in this work is costly. It should be remembered that, although coated base sheets in standard hot-roofing specifications are sometimes desirable, they should be handled in much the same way. At least they have the advantage of a 350° to 400°F mopping to soften them before they are covered with felt.

Equipment Required

Equipment will depend on how the cements and coatings are delivered to the job (i.e., 5-gal pails, 45-gal drums, or by tanker). Equipment is available to spray directly from drums on the roof or ground or from warming kettles. Hydraulic or air-operated piston pumps transfer cements and coatings from tankers to the roof. Other pieces of equipment remove old gravel from a roof and blow back roofing granules for the final surfacing.

Equipment Required

There are many ways of moving cold-application felt to roofs, all power assisted as for other forms of roofing. Each roofing contractor will choose equipment to suit the size and type of buildings roofed or will lease it as required, as for instance a compressor to drive air or hydraulic pumps.

High-pressure hoses and spray-gun heads are required, as well as propane-fired hot rollers to make sure that good adhesion is obtained. If the roof deck is uneven, the hot rollers are useless.

19

Fire Resistance and Fire Ratings

Roof system design is concerned with fire originating inside a building or from outside neighboring property. The exterior fire threat involves the roof membrane first, the substrate second, and the structural frame third. Except for direct heat radiation to a steep roof, slope is an advantage since burning brands have a tendency to slide off. A flat roof is not subjected to radiated heat unless a burning building is immediately adjacent and higher. In this situation, the flat roof must also be resistant to burning debris. Because it is flat, it is easily flooded by fire hoses and kept well below the point of ignition. A gravel surfacing on all types of felt would probably put the roof in a class A category, but an asbestos felt roof with gravel is probably the best. In an insulated roof system the performance of the insulation under elevated temperatures bears scrutiny. Lightweight plastic foams are the least desirable. When the insulation is placed on top of the membrane, as in the protected membrane system, a concrete block ballast would be preferable to a loose gravel ballast. Very often a neighboring fire produces frantic activity on a flat roof, perhaps more hazardous to the roof than the fire, in which case a protected membrane system with concrete block ballast is ideal.

The internal fire threat is first to the support framing, second to the deck, and third to the vapor barrier, insulation, and roof membrane. If a fire is not brought under control quickly by a sprinkler system, light steel framing and decking or light wood framing and thin wood decking could ignite or collapse.

Fire Resistance and Fire Ratings

In the case of steel, the deck deforms under heat (about 1,000°F) and allows easily ignited bitumen to melt and run into the building. A gypsum board cover would delay the final break-through if mechanically fastened to the steel. Poured-in-place reinforced dense concrete integral with columns and beams or precast concrete decking on a precast concrete frame would offer better protection.

Another important factor is the size of an uncompartmented space and the weight of the combustible contents, from which can be computed the fire load in Btu's per square foot and the severity and approximate length of time the fire will burn.

Roofing materials may carry Underwriters Laboratories (UL) labels for class A, B, or C compliance. Materials and systems are classified for their ability to protect the roof deck from the effects of fire exposure from the exterior (360-018) or (360-019). UL also rates complete systems under Section 360-RO for the fire hazard and fire spread on the underside of the deck. This includes the roof deck, adhesives, vapor barrier, and insulation.

UL Section 360 R-13 designates systems and materials for wind uplift resistance, but other properties that are important in the overall performances of a roofing system have not been rated.

Complete details of the information available should be obtained annually from Underwriters Laboratories, Northbrook, Ill.

The Factory Mutual Engineering Corporation in Northwood, Mass., publishes each year loss-prevention data for fire-retardation and wind-storm resistance. The designer must resolve any differences between Underwriters Laboratories, Factory Mutual Engineering Corporation, and local building regulations. As a rule, materials are not tested or evaluated for performance as a waterproof membrane.

Additional information may be required from the National Fire Protection Association and the National Building Code (Section 400).

20

General Maintenance

The existence of a long-term roofing warranty or guarantee issued by a manufacturer, supposedly supported or backed by an insurance company or surety, often misleads building owners into believing that they have no responsibility for their own roof. A short-term (two to five years) guarantee issued by a roofing association, if it is backed by or based on competent inspection during application, has some value, partly because the members of the association are usually seriously involved in maintaining high standards of workmanship. The association is more easily reached in the event of problems arising with the roof than might be the case with a large manufacturing company and insurance company in another state.

An inspection of any roofing bond, guarantee, or warranty will show that the liability assumed is limited and perhaps reduces in value as the roof gets older. An owner should relate its value to the cost per square and the percentage of the overall cost of the roof. Flashings are not covered by a roofing guarantee, even when there is a flashing endorsement, and there are usually several other exemptions from liability. It is recommended that a building owner or manager work out a maintenance program with an independent roofing contractor who is not involved in any way with a materials manufacturer and who is a member in good standing of a national, city, or state roofing association. The building owner/manager may also employ the services of a roofing inspector or consultant, who will provide an unbiased report on a roof's condition together with a specification of work that may be required.

General Maintenance

The following items should be checked at least once a year or more frequently if the weather is unusual or if the building is a factory, school, or any public building where roof traffic is possible, or any building in an industrial area. Many strange things can happen to flat roofs when no one is watching.

General Directions

When inspecting the roof, check carefully for

- Soft spongy areas
- Wet insulation
- Air-filled blisters
- Buckling, ridging, and roof cracking
- Cracks or splits due to shrinkage of roof membrane or movement in the deck
- Curled felt edges
- Wind-eroded gravel
- Exposed roofing felt
- Sliding of felt and gravel on inclines
- Loose or curled felt on smooth-surfaced roofs
- Serious alligatoring of top coating on smooth-surfaced roofs
- Shrinkage, delamination, and slippage on mineral-surfaced selvage-edge roofing
- Accumulation of dirt, fly ash, sawdust, and waste products from industrial plants
- Growth of moss and other vegetation
- Bottles, cans, stones, nails, waste wood, bricks, old signs, sea shells
- Poor installation of signs, mechanical equipment, television and radio antennas
- Exposed roofing at cant strips, below metal counter flashing
- Cracked roofing over end joints of gravel stops
- Blocked drains in roof and scuppers in parapets
- Loss of drain strainers
- Raised drains due to deck settlement or poor expansion sleeves
- Unusual deflection or settlement
- Loose and rusted flashings
- Low flashings on ponded roofs
- Empty gum pockets or pitch pans
- Loose and empty caulked counter flashings
- Loose reglet flashings
- Damage to expansion or control joints
- Damaged or blocked gutters and conductor pipes
- Cracked skylight glass and defective plastic domes

Need for traffic walkways

Damage to roof around mechanical equipment

Ponded water — check for chemicals

Fixed ladder access to all roof areas

Hatchway access to roof areas — should be easy, safe, and lockable (many are not and discourage regular inspection)

Only a few of the faults cited above can be repaired successfully with cold asphalt or coal-tar roofing cement or gum and perhaps a glass fabric reinforcement, and these repairs are usually only temporary. In wet cold weather they are usually ineffective because of surface contamination and poor bonding. A minor loss of gravel cover can be remedied by a heavy coating of cold cut-back, or in dry warm weather, an asphalt emulsion and fresh gravel; however, the use of hot material is infinitely better.

Major defects in the original roofing workmanship or building design require a professional roofer or sheet-metal mechanic, and often a completely different roofing system. Stop-gap repairs by in-house labor are sometimes unavoidable, but they are apt to become a habit because they offer temporary relief. In the long run, however, they add up to a relatively expensive repair policy and one that can hide structural damage until it is too late.

Another possibility is the entry of alleged roof repair experts from another city or state who will try to load you up with "magic" materials and a spurious guarantee.

Faults that arise out of ordinary wear and tear from the elements (covered by a guarantee or roof bond) usually can be repaired, but those arising out of internal defects in the system can not. Materials used in repairs should be similar to and compatible with the original; it is, therefore, essential that existing materials be properly identified. This is particularly important with caulking compounds and sealants. It is advisable to follow the manufacturer's directions and understand the mechanics of proper installation. Otherwise, a great deal of money can be wasted.

Summary

Many of the above items fall into the category of simple housekeeping chores, but the more serious faults can be due to several things that after a few years are often difficult to analyze. The division of responsibility for some failures is difficult to establish unless the roof is only one or two years old and all the facts are available for analysis by a competent roofing authority. Unfortunately, a roofing failure can end up in great confusion because of the many people who are directly and indirectly involved.

Studies of written judgments handed down in lower courts and appeal decisions can be useful because they often reveal weaknesses in certain roof systems or their components. They also show the importance of written instructions and decisions made during a roofing application and records of discussions between parties involved in the construction. In a court of law the importance of credible expert witnesses will be apparent. Finally, it is interesting to observe how important technical points are sometimes neglected or misunderstood by legal counsel and judges. A courtroom is not the best place to have a roof repaired or replaced. Court judgments are usually only sought when a roofing guarantee is involved.

21

Guarantees and Life Expectancy

Long-term guarantees for built-up roofs are not available from all manufacturers or in all parts of the United States. They are not available in any part of Canada. It is suggested that anyone who is considering purchasing or specifying any form of roofing guarantee first obtain a specimen copy and have it explained in detail by the guarantor in the presence of legal counsel and/or a roofing consultant.

One manufacturer of roofing materials and thermal insulation refers to its document in the literature as a Bonded Roofing Agreement, but does not include this wording on the document itself. In small print it is called a Certificate of Coverage. The manufacturer also states it is not a legal document, though it is signed by the vice presidents of three companies. The term of the bond or certificate and the maximum amount of dollar liability are typed in when issued. The name of the roofing contractor does not appear.

It is stated in the literature, but not on the certificate itself, that the terms and conditions of the Certificate of Coverage are subject to change without notice. This requires legal interpretation.

Flashing endorsements are available, but they do not include any metal, cap flashings or counter flashing, gravel stops, or edging. This is normal with all roofing manufacturers.

Guarantees and Life Expectancy

Owing to the many variables that affect the result, it is not possible to accurately predict how long a hot-applied built-up roof will last or what repairs may be required during its lifetime. Premature failure can be averted and repairs minimized by observing the various general rules and recommendations.

While a roofing manufacturer's bond agreement might, in some instances, be considered a sales tool, it is nevertheless a legal document that accepts responsibility for the quality of certain materials for as long as 25 years. Regardless of the value of such a document, it has more or less led people to believe that a roof life of that length is attainable. Experience shows that 25 years is realistic even in extreme climates of the North American continent, but the state of the art has not been stabilized to the point where it can be guaranteed every time. Unfortunately, laboratory testing cannot duplicate the forces of nature on a complete roofing system. Weatherometer testing of individual components is useful to a manufacturer, but the results require intelligent analysis and careful, responsible application to roofing in general.

Judging from what information is available and after considering the technical aspects, it seems reasonable that if the roof membrane is protected by insulation and ballast it will last longer.

The ever-changing environment will no doubt have an unknown effect on roofing materials and metal flashings. If more coal is used to generate electric power, we may have an increase in airborne materials such as soot, tarry matter, dust particles, various vapors and gases including carbon monoxide and carbon dioxide, water vapor, sulfur dioxide and oxides of nitrogen, as well as organic compounds, particularly hydrocarbons.

The burning of other fuels such as oil and natural gas similarly produces soot, vapors, and gases including sulfur dioxide, ammonia, methane, and acetylene, plus organic materials.

To illustrate the absurdity and uncertainty of some roofing guarantees, the following account of a roof applied on a steel deck in 1957 is presented. The information is from the summations of a justice of the high court of Ontario in 1962 (five years after the roof was laid), and the Supreme Court of Canada in 1966 (nine years after).

The roofing contractor first won a provincial court decision in an action brought by the manufacturers of steel products, but the decision was overturned by the Supreme Court of Canada on the basis of a different interpretation of the word "weathertight," and other points of law.

A cold ahdesive (Curadex), probably a chlorinated rubber, used to fasten perlite (J.M. Fesco board) insulation to a sloped steel roof deck, did not hold the insulation to the steel in three violent wind storms. The adhesive was preferred by the owner and manufacturer of the steel deck to a mechanical fastener because it was cheaper and eliminated perforations of the deck. The use of adhesive was made possible because there was no air/vapor barrier. The deck was steep enough to require wood ledger strips at 6'-0" on center (o.c.) from eaves to ridge to prevent sliding.

It was unlikely the insulation could stay in place under a high wind suction without mechanical fastening. In the area where the buildings were located, it was a well-known fact that wind gust speeds approaching 90 mph could be expected two or three times a year. In this particular case, the four-ply smooth-surfaced asphalt felt roof and insulation were removed by wind on three separate occasions. When the roofs were replaced, the insulation, presumably the same type, was secured with Riv-Nails (see Figure 4-6).

The negative pressures or uplift forces at these speeds are roughly 16 lb/ft^2. In view of the small area of adhesive applied in a serpentine pattern, the negative force per square inch of adhesive could be very high. The roof and insulation weighed only 2.75 lb/ft^2, and the deck about 2 lb, therefore there was insufficient dead load to resist the uplift forces or prevent fluttering of the deck, and the disturbance of all the components in the system.

Perlite board is made with expanded mineral pellets (volcanic ash), wood fiber, and asphalt binder. It is highly absorbent, is quite friable, and has a low horizontal shear strength. In other words, it is not suitable for use over steel decks without mechanical fastening or where there is liable to be a high wind suction. In high winds, the deck fluttering or vibration could easily break the bond between the insulation and the steel deck by fracturing the perlite or the adhesive.

The owner drew up the roofing specifications and made it a condition of the roofing contract that the roofer would guarantee the roof to remain "weathertight" for five years. The Supreme Court said "weathertight" meant the roof specifically had to resist both usual and unusual winds. It is worth mentioning that a J.M. roof guarantee at the time denied responsibility when wind velocities exceeded 47 mph. Most manufacturers' roof bonds refuse responsibility for damage by wind of any velocity. If the word "weathertight" had been changed to "watertight," it might have been a different ball game. Even without a roof membrane, a sloping steel deck with vertical flutes would not leak. As it was, there was no complaint of leakage, only that parts of the roof and insulation had blown off, even if they were not needed. Steel deck manufacturers do not guarantee that their products will not cause any problems in a roof system, and yet in the roofing industry they are one of the principal headaches because special precautions must be taken that are not required with other roof decks. Even with the necessary precautions they are not the best base for a roof. It is interesting to note that the steel deck manufacturer did not question the suitability of his own product and neither did anyone else. The judge of the lower court did say that from the evidence the defendent corporation was taking a calculated risk in specifying the adhesive designed and required to fasten the roofing membrane to a roof of new design (sic). (The wording of the judgment is often puzzling and indicates a fuzzy understanding of roofing systems.) The judge continues by saying that the defendant corporation knew this to be the case. The manufacturers of the adhesive and the insulation, surprisingly, were not made a party to the suit against the roofing contractor and yet it was the interaction of the steel roof deck, the adhesive, and the insulation that resulted in the loss of the roof. No blame was given to the workmanship of the roofing contractor by anyone and this was recognized by the court.

The buildings involved are identified as the Roll Shop and Pickler Aisle. In hindsight, it can be argued that the whole situation could have been avoided if the steel deck had been designed to serve as the water shedding roof, and the redundant insulation substrate and the felt and asphalt roof omitted. There was adequate slope for drainage and it is doubtful that thermal insulation was needed in the system. There was no air/vapor barrier and the perlite, assuming a 1-in. thickness, only provided a low resistance of 2.78, with only 4.0 for the whole system, including air

Guarantees and Life Expectancy

films. In addition to all this, the smooth-surfaced asphalt roof is costly to maintain.

Fesco board was probably selected for its alleged fire resistance, although the roof covering is only Class C.

The judgment cost the roofing contractor nearly $15,000 plus interest of about $3,000, together with costs of the action. The roofing contractor was held responsible for a poorly designed roofing system, not a poorly applied roof membrane. The contractor completed the work to the complete satisfaction of all concerned but lost the case because of one word in the guarantee.

One would think that the steel deck manufacturer would recognize and admit his mistake and modify the recommendations for the safe use of his product. The modifications, however, eventually had to be made by Factory Mutual Engineering Corporation. (See the section on steel decks in chapter 4.)

22

Sources of Technical Literature

Abraham, Herbert. *Asphalt and Allied Substances*, 6th ed. New York: Van Nostrand Reinhold Publishing, Co., 1960.

American Iron and Steel Institute, 1000 16th St., N.W., Washington, D.C. 20036

American Plywood Association, Applied Research Department/Technical Services Division, Tacoma, Wa. 98401

American Society for Testing and Materials (ASTM), 1916 Race St., Philadelphia, Pa. 19103

Asphalt Roofing Manufacturers Association, 1800 Massachusetts Ave. N.W., Suite 702, Washington, D.C. 20036

Atmospheric Environment Service, 4905 Dufferin St., Downsview, Ontario, Canada M3H 5T4

Baker, Maxwell C. *Roofs: design, application and maintenance.* Montreal, Quebec, Canada: Multiscience Publications Limited, 1980, 361 pp.

Canadian General Standards Board (Sales), Supply and Services Dept., Phase III-4B1, 11 Laurier Street, Hull, Quebec, Canada K1A 0S5

Canadian Roofing Contractors Association, Ottawa, Ontario, Canada K1P 5G3

Canadian Sheet Steel Building Institute, Willowdale, Ontario, Canada M2J 4G8

Sources of Technical Literature

Factory Mutual Engineering Corporation, Norwood, Mass. 02062

Griffin, C.W. Jr., P.E. *Manual of Built-up Roof Systems*, 2nd ed. New York: McGraw-Hill Book Company, 1982.

National Bureau of Standards, Washington, D.C. 20234

National Research Council — Division of Building Research, Ottawa, Ontario, Canada K1A 0R6

National Roofing Contractors Association, 8600 Bryn Mawr Avenue, Chicago, Ill. 60631

Pennsylvania State University College of Engineering, University Park, Pa.

Portland Cement Association. *Design and Control of Concrete Mixes*, Skokie, Ill. 60076; or 116 Albert Street, Ottawa, Ontario, Canada K1P 5G3

"Proceedings of the Symposium on Roofing Technology," sponsored by the National Bureau of Standards and the National Roofing Contractors Association, September 1977.

Rogers, Tyler Stewart. *Thermal Design of Buildings*. New York: John Wiley & Sons, Inc., 1964.

Roofing Industry Educational Institute (RIEI), 6851 S. Holly Circle, Suite 250, Englewood, Colo. 80112

Sheet Metal and Air Conditioning Contractors National Association. *Architectural Sheet Metal Manual*, May 1968, 2nd ed. Washington, D.C.

Sweets Architectural Catalog File

Thomas Register Catalog File

Underwriters Laboratories, Northbrook, Ill.

Underwriters Laboratories of Canada, 7 Crouse Road, Scarborough, Ontario, Canada M1R 3A9

University of Minnesota Institute of Technology, Minneapolis, Minn.

University of Illinois — Small Homes Council, Urbana, Ill.

U.S. Department of Commerce — Environmental Data Service, National Climatic Center, Asheville, N.C. 28801

23

A Glossary of Roofing Terms*

Adhesive: A cementing substance that produces a steady and firm attachment between two surfaces.

Air Barrier: A membrane or building element that provides resistance to air leakage.

Air Leakage: The movement of air through spaces between constituent parts of a roof system or other enclosure element as a result of air-pressure differences between one and the other side.

---

* Glossary of roofing terms by the kind permission of Mr. Maxwell C. Baker and the National Research Council, Canada. Extracted from *Roofs* by Maxwell C. Baker, Multiscience Publications Limited, Montreal, P.Q. 1980.
Note: The inclusion of the Glossary from *Roofs* does not in any way imply approval of any statements made in Commercial Roofing Systems.

Air Space: A cavity or unfilled space between two constituent parts in a roof system or other enclosure element of a building.

Aggregate: Material such as natural gravel, small-size broken stone or crushed slag, used as a protective surfacing or ballast in roofing systems.

Alligatoring: Hardening and shrinking of exposed bitumen coatings due to oxidation, that produces small islands of bitumen between deep cracks, the overall appearance of which somewhat resembles alligator hide.

Application: The act of putting on or building up the felts and flashings of a built-up roofing, or all the elements of any roofing system.

A Glossary of Roofing Terms

Cold Applications: The applying of felts in a built-up roofing with cold bituminous cements.

Horizontal Application: Mineral-surfaced roofing applied with the laps parallel to the eaves of a sloping roof.

Hot Application: The application of felts in built-up roofing using heated bitumen.

Phased Application: The practice of laying one or more plies of a built-up roofing at one time with the additional plies laid at a later time.

Separate Layer Application: Felts applied with a small edge-lap for each of several separate plies.

Shingle Application: Felts applied in an overlapping manner similar to shingle application, with the amount of overlap arranged to give the number of plies desired.

Two-and-Two Application: A four-ply roofing laid in shingle fashion with the first two giving double coverage and the last two separate double coverage.

Vertical Application: Mineral-surfaced felt applied with the laps at right-angles to the eaves and parallel to the rake. Also called up-and-over when it continues over the ridge. Sometimes laid slightly on the bias to encourage drainage away from the laps.

Asphalt: Dark brown to black hydrocarbon solids or semi-solids that gradually liquefy when heated. It occurs in natural deposits at a few places throughout the world, but the source of asphalt for roofing is the residuum of petroleum distillation.

Attic: A roof space between the top floor ceiling and the roof of a building.

Barge Board: A board, often decorative, covering the projecting portion of a gable roof.

Base Sheet: A heavy sheet of saturated or coated felt placed as the first ply in a built-up roofing membrane of roof system.

Bitumen: A generic term applied to mixtures of predominantly hydrocarbons in viscous or solid form, derived from coal or petroleum. The roofing industry uses it to describe either coal tar pitch or asphalt.

Blister: An enclosed raised spot or area containing gas or liquid that shows at the surface of prepared or built-up roofing.

Surface Blisters: Small blisters from pin head size to usually less than 25 mm in diameter appearing in the surface coatings of roofing. They frequently occur in clusters and result from exposure to sunlight and weather. Also called weather blisters, pin blisters, blueberries, pimpling, and bitumen bubbling.

Structural Blisters: Blisters between plies of felt in a built-up roofing membrane, or between the membrane and its substrate, ranging in size from 25 mm dia. and barely visible height to 5 m² in area and 300 mm in height. Also called interply or interface blisters.

Blocking: Continuous strips, usually of wood, secured to roof decks at the perimeter edges and around roof openings to provide securement for the roofing membrane and flashings, or for other building parts.

Bond: 1. To hold together two roofing components by means of an adhesive. 2. The adhesive strength that prevents delamination of two components. 3. A guarantee relating to roofing performance.

Brooming: The pressing of felts in close contact with the layer of bitumen immediately following the application of bitumen and felt, by the use of a wide stable of deck-type broom or other suitable push bar as wide as the felt.

Buckle: Large elongated bulge or fold in a roofing membrane as a result of separation from the substrate accompanied by expansion or stretching.

Cant: A continuous strip of material of triangular section placed at the intersection of a roof deck with a higher wall or other vertical surface. The roofing membrane and flashings are eased through the change in direction from essentially horizontal to vertical along its 45° sloping surface.

Cap Sheet: 1. The top ply of a built-up roofing membrane acting as the finished surface of a roof. 2. Any mineral-surfaced or other coated felt or sheet designed for that purpose.

Caulking: Any of a wide range of bituminous, rubber, plastic or other materials suitable for filling seams or cracks to make them tight against water leakage.

Cement: A substance used to make objects adhere to each other. In the roofing industry loosely applied to mean caulking and mastic.

>*Flashing Cement:* A trowelable mixture of asphalt, volatile solvent and mineral fillers used as a cold coating in the application of flashing, for sealing around roofing details and for cold patching.
>
>*Plastic Cement:* Same as flashing cement.
>
>*Portland Cement:* Hydraulic cement used for making concrete and grout.

Coal Tar: Tar derived from the destructive distillation of coal during the conversion of coal into coke.

Coal Tar Pitch: A dark brown to black hydrocarbon solid or semi-solid specifically refined for roofing from coal tar.

Coating: A thin layer of a substance used to cover other materials, to provide an aesthetic or protective function.

Cold Process: 1. Roofing comprised of layers of bituminous-coated felt adhered with cold-applied bituminous cement, and surfaced with emulsion or cut-back. 2. Sometimes applied to any roofing system that uses bituminous materials applied cold.

Collar: A metal cap flashing around a vent pipe projecting above a roof deck.

Condensation: The change from water vapour to liquid water, resulting from a drop in temperature of an air vapour mixture.

>*Concealed Condensation:* That which takes place within a roofing system and is not seen.
>
>*Interstitial Condensation:* That which occurs in the interstices between constituent parts of a roof system. Same as concealed.
>
>*Surface Condensation:* That which appears on the colder exposed surfaces of a roofing system.

Conductor: A pipe for conveying rain water from a roof gutter to a drain, or from a roof drain to a

A Glossary of Roofing Terms

storm drain. Also called a leader, down spout or down pipe.

Control-flow: Relating to roof drainage. A type of drain or a system of drains that regulates the flow of water so that rain water can be drained away at a uniform rate no matter how heavy the rainfall.

Coping: The cap or highest covering course of a wall, usually overhanging the wall and having a sloping top to shed water.

Cornice: Projection at the top of a wall. Term applied to construction under the eaves where the roof and side walls meet. The top course, or courses of a wall when treated as a projecting crowning member.

Coverage: The surface area covered by a specific unit of roofing material after allowance is made for the required edge lap or overlap.

Crack: A break in a roofing membrane as a result of flexing, often at a ridge or wrinkle.

Crazing: Surface deterioration by the formation of a pattern of fine hairline cracks.

Cricket: A small false roof, or the elevation of a part of a roof surface, as a means of diverting water from behind a projection such as a chimney. Also used to direct water to drains in a horizontal roof valley formed by the intersection of two sloping roofs. Also called a saddle.

Curb: A low wall of wood or masonry built above the level of the roof, surrounding a roof opening such as is required for installation of fans and other equipment, and at the edges of movement joints in a roof deck.

Cutback: A solution of bitumen in a volatile solvent used as a primer, cold-application cementing agent or roof coating. Filled cutbacks may contain mineral particles and inorganic fibres.

Cut-off: A detail intended to prevent lateral transfer of water in the insulation of a roofing system. A membrane placed along joints to separate roof insulation into multiple areas, or turned over the edges to protect insulation at the roof perimeter, or to seal the edge of insulation at the end of a day's work during roofing application.

Dampproofing: The treatment of a building material or component surface with a bituminous or other coating to provide some measure of resistance to the passage of moisture into or through the material or component.

Dead Level: A roof deck with no intentional slope to the roof drains.

Deck: The structural infill between main structural supports, to the top surface of which a roofing system is applied.

Delamination: Separation of the felt plies in a built-up roofing membrane.

Dew Point: The temperature at which a sample of humid air, cooled from some temperature, becomes saturated and at which water vapour begins to condense to liquid water.

Dipper: A ladle for pouring bitumen.

Dormer: A separate smaller roofed structure that projects from a sloping roof to provide more space below the roof and to accommodate a vertical window.

Double Pour: The application of the top covering of bitumen and gravel surfacing of a built-up roofing in two separate operations. A quantity of gravel is spread over a first-pour coat of bitumen, loose gravel is removed, and additional gravel is spread into a second-pour coat of bitumen.

Downspout: See Conductor.

Drain: An outlet to allow water to flow from a roof surface into a drain pipe and away from the building through a drainage system.

Dip Edge: The formed edge on metal flashing used at the eaves or other roof details to encourage water to drip away from vertical surfaces of the building detail.

Drippage: Bitumen that flows and drips through holes or over the edge of a roof deck.

Dry Laid: Any roofing felt laid without bitumen or other adhesive.

Duck Boards: Slatted wood-board panels for placement on a roof to provide a walkway or roof surfacing for light traffic.

Edge Venting: The practice of providing regular spaced or continuous openings at a roof perimeter to relieve water vapour pressure or dry out the roofing system, usually combined with venting channels in the insulation and stack venting towards the centre of the roof.

Emulsion: An intimate mixture of fine globules of bitumen held in suspension in water by means of a chemical or clay emulsifying agent.

Equilibrium Moisture Content: The balanced moisture content attained by a material at any particular temperature and humidity conditions expressed as a percentage of moisture mass to material mass.

Equiviscous Temperature: The temperature at which bitumens will have the correct viscosity for spreading in roofing application.

Expansion Joint: A deliberate joint separation of two parts of a building through floors, walls and roof to allow expansion and contraction movement of the parts. The joint is provided with a flexible watertight connecting detail.

Exposure: The amount of any particular roofing unit exposed to the weather or not covered by an overlapping unit in a roofing system that utilizes overlapping application. The dimension describing this is measured in the direction of the overlap and is normally the unit width minus a small amount to ensure complete coverage, divided by the number of plies. For a two-ply membrane from 1 m wide roofing felt allowing 50 mm to ensure coverage the exposure would be $(1000 - 50)/2 = 475$ mm. For shingle type roofing the allowance is called the headlap.

Fabric: A woven cloth of organic or inorganic fibres treated with bitumen and used for special flashing applications.

Fall: The vertical distance in millimetres through which a roof incline falls in a unit horizontal distance of one metre.

Fallback: A reduction of bitumen softening point related to contamination, incompatibility or over heating. Also referred to as softening-point drift.

Fascia: The finish member covering the edge or eaves of a flat or sloping overhanging roof.

Feather: To reduce the edge of a material to a very small dimension like a feather edge.

Felt: A general term used to describe sheet roofing material consisting of a mat of organic or inorganic fibres

A Glossary of Roofing Terms

untreated, or saturated, or saturated and coated with bitumen and supplied for use in roll form.

Asbestos Felt: Felt containing from 75% to 85% of asbestos fibre.

Asphalt Felt: Felt for which the bituminous saturant or coating is asphalt.

Coated Felt: Asphalt-saturated felt coated on one or both sides with filled asphalt.

Dry Felt: Organic-fibre roofing felt before any treatment with bitumen. Used as an underlayment for built-up roofing over wood-board decks to prevent bitumen drippage or to provide a slip sheet.

Glass Felt: Felt made from glass fibre.

Mineral-Surfaced Felt: Bitumen-coated felt surfaced on one side with natural or synthetic coloured granules.

Organic Felt: Felt made from organic fibres and in particular wood fibres.

Perforated Felt: Bitumen-saturated felt perforated with closely-spaced small holes to allow for escape of air and moisture during application.

Rag Felt: A term sometimes used to describe organic-fibre felt. A hangover from earlier days when a percentage of rag fibre was used.

Saturated Felt: Felt which has been impregnated with bitumen by passing it through vats of hot saturant.

Stripping Felt: Narrow widths of felt used to complete flashing details, particularly to cover the edges of metal flanges incorporated into built-up roofing.

Tar Felt: Felt for which the saturant is coal tar pitch, more properly called coal tar pitch felt.

Felt Layer: A piece of mobile mechanized roofing equipment for spreading bitumen and laying felt in a single continuous operation.

Fill: Aggregate and cement mixtures placed on a roof deck in varying thickness to level out depressions and irregularities, or to form slopes to roof drains.

Filler: Finely-divided mineral matter used as an extender and to improve the properties of asphalt coatings for shingle and built-up roofing felts, and bituminous-plastic cement or mastic. Also called a stabilizer.

Firewall: Any wall built for the purpose of restricting the spread of fire in a building. Such walls of solid masonry or concrete usually divide a building from the foundations to about a metre above the roof.

Fishmouth: An opening occurring at the lapped edge of applied felts in built-up roofing because of adhesion failure. May be isolated occurrences or in a more or less regular pattern.

Flashing: A building device used to prevent water from penetrating the exterior surface of a building element or material, or to intercept and lead water out of it. Flashing can be considered as a continuation of the roofing membrane to protect and weatherproof any element of the building or roof deck that departs from the roof deck level or incline.

Base Flashing: The extension over a cant strip and up the vertical surface, of the roofing membrane at the base of a vertical wall or item intersecting or penetrating the roof.

Cap Flashing: The sheet-metal coping for the top of a higher wall such as a parapet, or the cover over a detail such as expansion joint.

Counter Flashing: The material, usually sheet metal, protecting the top edge and covering or partially covering the base flashing. Sometimes also called a cap flashing.

Eaves Flashing: The treatment of the edge of a roof with felt and metal flashing. The portion of the metal eaves flashing exposed on the elevation may be called a fascia flashing.

Gravel Stop: A formed strip of metal at the edges of a gravel-surfaced roof to prevent the gravel from rolling or washing off. Usually combined with the eaves flashing to add a crisp finished appearance to the roof edge.

Step Flashing: Individual pieces of flashing material used to counterflash chimneys, dormers and such projections along steep-sloping roofs. The individual pieces are overlapped and stepped up the vertical surface.

Through-the-Wall (or Thruwall) Flashing: Flashing extending completely through a masonry wall to lead water that penetrates higher up out of the wall at the flashing.

Flood Coat: The top layer of bitumen for an aggregate-surfaced built-up roofing membrane, poured or flooded onto the finished felts and over which the aggregate is spread. Also called a pour coat.

Gable: The triangular end of an exterior wall from the level of the eaves to the ridge of a double-sloped roof. A gable roof is a ridged or double-sloped roof which terminates at one or both ends in a gable. A gable end is the end wall of a building with a gable formed by the roof.

Gambrel: A type of roof which has its slope broken by an obtuse angle, so that the lower slope is steeper than the upper slope. A double-sloped roof having two inclines on each slope.

Glaze Coat: 1. A thin coating of bitumen applied to the felts of unfinished roofing to give short time protection from weather when roofing operations are delayed. 2. Also refers to the top layer of asphalt in a smooth-surfaced built-up roofing.

Granules: Mineral particles of a graded size that are embedded in the coating asphalt of shingles and mineral-surfaced roofing.

Gravel: Small pieces of aggregate larger than sand grains resulting from the natural erosion or the crushing of rock used as a protective surfacing or ballast in roofing systems.

Gravel Spreader: A piece of mobile mechanical roofing equipment that dispenses bitumen and spreads gravel in one continuous operation.

Gravel Stop: The upward projecting edge of an eaves flashing to stop gravel from rolling or being washed off from an aggregate-surfaced built-up roofing.

Gravelling In: The operation of spreading a gravel surfacing over the flood coat of a bituminous built-up roofing.

Grout: A fluid cement-mortar mixture

A Glossary of Roofing Terms

used to fill joints and cavities of masonry or concrete building construction. On roof decks the joints between many types of precast roof deck slabs are grouted.

Gutter: Trough at the eaves of a roof to convey rain water from the roof to a downspout.

Header: The beam into which the common joists are fitted when framing around a roof opening. The headers are placed so as to fit between two long beams or trimmers to support the joists ends.

Hip: The sloping line along the outer angle formed by the meeting of two sloping sides of a roof whose eaves meet at a right-angle. A hip roof is one that rises by inclined planes from all four sides of a building to form hips at the intersection of adjacent roof slopes.

Hood: A sheet-metal cover over equipment, stack vents or similar roof details.

Hygroscopic: Attracting and absorbing moisture from the air.

Ice Dam: Ice formation at the eaves of snow-covered sloping roofs that forms an obstruction to the drainage down the slope of snow-melt water.

Incline: The angle made by a roof plane with a horizontal plane. Interchangeable with slope or fall.

Insulation: A material used as part of a building enclosure to retard the flow of heat through the enclosure.

Jack: A flanged metal sleeve used as part of the flashing around small items that penetrate a roof.

Joist: One of a number of the smaller closely-spaced parallel structural supports for a flat roof-deck spanning between walls, roof beams or purlins, or to support a flat ceiling below a sloping roof.

Kettle: Equipment used for heating bitumen to the temperatures required for application.

Kettle Temperature: The temperature to which bitumen is heated in the roofing kettle, often considerably higher than that at the point of application.

Kettle Thermometer: A thermometer used for checking the temperature of the heated bitumen in the kettle.

Lap: That part of a roofing unit that covers the preceding course in any overlapping roofing application. Applied to shingles, built-up roofing felts, and most other types of roofing.

Edge Lap: The amount of overlap of the edge of a ply of roofing felt over the previous ply. Also called the side lap.

End Lap: The amount of overlap at the start of a roll of felt over the end of the previously laid roll.

Head Lap: In shingle or other overlapped unit roofing the amount that the head of an underlying unit is lapped by the lower edge of the uppermost overlying unit at that location. For double-coverage units the head lap is the unit width minus twice the exposure.

Lap Cement: A cut-back asphalt used for cementing the overlaps of cold-application roll roofing.

Leaching: The dissolving-out of soluble substances when water runs slowly through a roofing system, often responsible for ugly staining on ceilings and walls when the water drains to the interior.

Leader: Drain pipe, downspout or conductor. Also rain water leader abbreviated to RWL.

Leanto: A sloping roof resting against a higher wall of a building.

Mastic: Trowellable bituminous paste made by adding mineral fillers to concentrated cutbacks. Also called plastic or flashing cement.

> *Mastic Pan:* A flanged metal collar incorporated into a built-up roofing membrane around a penetrating item through the roof and filled with mastic.

Mansard: A roof which rises by inclined planes from all four sides of a rectangular building. Each sloping roof has two inclines, the lower one usually very steep and the upper one almost flat.

Membrane: A continuous flexible or semi-flexible roof covering that forms the water-control element of a roofing system. It is usually built-up on site from single or multiple plys of material, e.g. polyvinyl chloride roofing in single-ply and bituminous-felt roofing in multiple-ply.

Membrane Migration: Progressive movement of roofing membranes in one or in both directions that can occur on roofs due to thermal shrinkage. It can move improperly-adhered insulation and tear flashing at roof edges.

Mill Deck: A type of wood roof-deck constructed from wood planks placed on edge vertically and spiked or nailed together.

Mini Mopper: A small container with wheels that can be pushed along over the roof to dispense bitumen for the laying of roofing felts.

Mop: A tool used for the application of hot bitumen made from a bundle of cotton or other yarn attached to a long wooden handle. Bitumen soaked up and held by it when dipped into a container of hot material is transferred to and spread on the roof.

Mopping: 1. The fact of spreading hot bitumen with a mop. 2. Also may refer to a layer of hot bitumen mopped between plies or over roofing felts.

> *Full Mopping:* Application to provide a continuous reasonably-uniform layer of bitumen over the entire surface being mopped. Also called solid mopping.
>
> *Spot Mopping:* Application of bitumen in roughly circular spots (400 mm to 500 mm in diameter) in a uniform pattern providing unmopped strips in a grid pattern or between staggered spots.
>
> *Strip Mopping:* Application of bitumen in parallel bands roughly 200 mm wide with 100 mm unmopped bands between. Also called channel or ribbon mopping.
>
> *Sprinkle Mopping:* Application by haphazardly sprinkling or dribbling of small amounts of bitumen onto a surface with a mop or broom. Also called drip or dribble mopping.

Movement Joint: See expansion joint.

Nailer: A member, usually of wood, set into or secured to nonnailable roof decks or walls to allow for positive anchorage by nailing of roofing felts, insulation or flashings. Also called nailing strips.

Nailing: Fastening of roofing materials by nails or other hammer-driven special fasteners.

> *Back Nailing:* The practice on sloping roofs of blindnailing overlapping roofing felts to a nailable substrate or to specially

A Glossary of Roofing Terms

provided nailing strips in addition to adhering all the plies with bitumen to prevent slippage.

Blind Nailing: Application of roofing in such a manner as to cover all nail heads by over-lapping material.

Concealed Nailing: Same as blind nailing.

Exposed Nailing: Application where the nail heads are exposed to the weather.

Overhang: The part of a roof structure that extends beyond the exterior walls of a building.

Overheating: The heating of bitumen for application of roofing to a temperature that permanently alters the characteristics of the material.

Parapet: A low wall along the edge of and surrounding a roof deck. It is generally an extension of exterior building walls or party and fire walls that usually extend about a metre or less above the roof.

Parting Agent: Fine sand, mica, talc or similar material spread over the surface of coated bituminous felt to prevent sticking in the roll.

Penetration: A measure of the hardness related to viscosity of bitumen as determined by an empirical test that gives the depth of penetration of a weighted needle into a sample after a definite time and at a particular temperature.

Perlite: An aggregate used in light-weight concrete and preformed insulating board, produced by heating and expanding silicaceous volcanic glass.

Pitch: A black or dark brown solid cemetitious residue that results from the distillation of tar. A tar derived from coal is referred to as coal tar, and a pitch derived from coal tar as coal tar pitch.

Pitch Pocket: A flanged metal collar placed over penetrating items on roofing and filled with coal tar pitch. Plastic or mastic pans also sometimes called pitch pockets.

Roofer's Pitch: Coal tar pitch.

Pimpling: See blisters, surface.

Plank Deck: Wood deck of planks usually 40 mm to 90 mm thick and 150 mm to 200 mm wide laid on the flat with tongued-and-grooved or splined edges, and spiked together.

Plaza: A roof terrace.

Ply: A single layer or thickness of roofing material in a roofing membrane.

Podium: A roof terrace.

Ponding: The collection of water in shallow pools on the top surface of roofing. This is generally from rain, but certain roofs are designed to hold a shallow depth of water over the whole roof surface for evaporative cooling in summer often with a water supply to the roof.

Pour: A layer of bitumen deposited on the roof surface or the felts by pouring from a bitumen container.

Double Pour: The application of the top layer of bitumen and the gravel-surfacing of a built-up roofing in two separate operations. This is accomplished by spreading gravel into a first pour coat, brooming off the loose gravel, and then applying additional gravel to a second pour coat.

Pour Coat: Same as pour.

Top Pour: The application by

pouring of the top layer of bitumen on a built-up roofing. Often used to describe the top layer of bitumen no matter how applied.

Primer: A thin solution of a coating applied to a surface to improve the adhesion of a heavier coating. Usually refers to a cutback bituminous coating of thin consistency.

Promenade: A roof terrace.

Protected Membrane: A roofing membrane with insulation and protective surfacing or landscaping outward from it. Also called inverted or upside-down roof.

Purlin: A horizontal structural member spanning between beams, frames or trusses to support a roof deck or the rafters or joists supporting a roof deck.

Rafter: One of a number of closely spaced structural members of a sloped roof, usually extending from the eaves to a ridge or hip on a small roof or between purlins on larger roofs to carry the roof deck.

Rake: The edge of a roof at its intersection with a gable.

Reglet: A horizontal groove or slot in a wall or other vertical surface projecting above a roof surface into which flashing can be secured and sealed.

Relative Humidity: A ratio expressed as a percentage of the mass of water vapour present in an air vapour mixture and the mass of water vapour that would be present in the sample of air if it were saturated at the same temperature. It can be stated also as a good approximation that it is the ratio of the vapour pressure of water present in a sample of air to the saturation vapour pressure at the same temperature.

Ridge: The horizontal line where two opposite sloping sides of a roof join at the highest level of the roof.

Ridge Board: A horizontal board in wood frame construction at the upper end of the common rafters to which the rafters are nailed.

Ridge Cap: The covering of wood, metal or other roofing material that tops the ridge of a roof.

Ridge Course: The last or top course of roll roofing, shingles or tiles on a sloping roof cut to length as required.

Ridging: A roofing defect characterized by narrow or relatively narrow ripples in a membrane generally along the machine direction for roofing felts and over deck or insulation joints and usually less than 25 mm in height.

Roof: A construction on top of a building that together with walls forms a separator between inside and outside environments. A roof system is a structurally supported, air, heat, interior moisture and rain control combination.

Dead Flat Roof: No intentional slope.

Extra Steep Roof: Slope over 1:1 (45°).

Flat Roof: Slope from 1:50 to 1:6 (1° to 10°).

Steep Roof: Slope from 1:6 to 1:1 (10° to 45°).

Roof Coating: A thin layer of filled bitumen applied to saturated felt. Filled or unfilled bitumen cutback, asphalt emulsion, or a compatible paint applied in a thin layer to pro-

A Glossary of Roofing Terms

vide a protective cover for roofing materials.

Roof-Deck: The structural infill between structural supports which forms the load carrying base for the rest of the roofing system.

Roof Divider: A building detail used to limit the size of a continuous roof membrane, dividing a roof into a number of smaller areas. The divider extends only to the roof deck and is not an expansion joint.

Roof Drain: The termination or fitting at the roof of an interior drain pipe or leader for draining water from a roof.

Roof Guard: A contrivance fitted to a steep sloping roof to prevent the sliding of snow or ice.

Roof Insulation: Any medium or low-density material suitable and used as part of a roofing system to reduce heat loss or gain through the roof.

Roof Terminal: The upper end of a vent stack above a roof. Drains also sometimes called terminals.

Roof Terrace: A traffic-bearing or landscaped roof. Also called promenade, podium or plaza-deck roofs or roof gardens.

Roofing: 1. The material used for constructing a water-shedding or waterproofed roof. 2. That part of the architectural specifications and building construction contract that deals with the supply and application of roofing materials and systems.

Built-Up Roofing: A continuous, semi-flexible roof covering built-up on site from alternate layers of bitumen and bitumen-saturated or coated felts often abbreviated to BUR.

Composition Roofing: All types of asphalt rolled roofing and shingles.

Mineral-Surfaced Roofing: Roofing that is coated on both sides with asphalt and finished on the weather side with natural or synthetic coloured mineral granules, usually for only part of the width of the felt.

Prepared Roofing: Same as composition.

Roll Roofing: Any roofing material which is supplied from the manufacturers in rolls, but more specifically applied to coated felts either smooth or mineral-surfaced used for roofing without additional top coatings or surfacings.

Smooth-Surface Roofing: Roofing felt that is asphalt-coated on both sides with either a smooth or veined surface. Built-up roofing that may have an applied coating but which has no protective surfacing of gravel or other aggregate.

Wide-Selvage Roofing: Mineral-surfaced roofing designed for double coverage in which the selvage is slightly greater than half the width of the felt.

Run: The horizontal distance to which the fall or vertical distance for an inclined roof is referenced. A unit horizontal distance of one metre is taken for the run to which the fall in millimetres is given to describe the incline.

Saddle: A ridge in a roof deck that divides two sloping parts of the surface so that water will be diverted to roof drains. Usually constructed in a level valley, or behind a projection above a sloping roof. Also called a cricket.

Saturant: A bitumen of low softening point for impregnating the dry felts in the manufacture of saturated roofing felts.

Saw Tooth: A roof formed by a number of north light trusses. When viewed from the end of the building such a roof presents a serrated or toothed profile.

Scraper: A tool or a piece of equipment for removing aggregate surfacing from built-up roofing for repair or re-roofing. Also called a spud or spudder.

Screed: Lightweight fill placed on the surface of a roof deck to create slopes to roof drains. Also the guide used to achieve the sloped fill.

Scupper: An outlet in the wall of a building or a parapet wall for drainage of overflow water from a floor or roof directly to the outside. Special scupper drains connected to internal drains are also sometimes installed at roof and wall junctions.

Scuttle: A small opening provided with a waterproof cover through the ceiling and roof to provide access to the roof from the interior. The scuttle may have its own curb, or may be placed on a built-up curb. Also called a roof hatch.

Seal: 1. A substance used to close a crack or other aperture against air or water leakage. 2. Narrow strips of bituminous material used to fill or cover such apertures.

Self-healing: Used in reference to bitumen that softens with heat from the sun and flows to seal cracks that earlier formed in the bitumen from other causes.

Selvage: The portion of mineral-surfaced roofing where the mineral surfacing is omitted to allow for the overlapping sheet to achieve better adhesion. For double-coverage application the selvage width is half the width of the roll plus about 25 mm and for single-coverage the roll width less 50 mm.

Sheathing: Board or sheet-type material fixed to studding or roof rafters or joists as the base for application of wall cladding or roof covering.

Sheathing Paper: A medium-to-heavyweight wood-fibre paper or felt often fastened to sheathing as the base for the application of exterior covering materials.

Shed: A roof having only one incline that slopes from a higher to a lower wall. A leanto roof sometimes also called a shed roof.

Shedding: The loss of mineral surfacing from prepared roofing.

Sheet: An unrolled piece of roofing felt.

Shingling: 1. The application of any roofing material by overlapping of units in horizontal courses with the overlapping down the slope to shed water. 2. The usual method of laying roofing felts in built-up roofing with overlapping sufficient to produce the number of plies desired.

Slag: A gray porous aggregate produced by air cooling and crushing residue from blast furnaces, used as a protective surfacing for built-up roofing.

Slippage: Sliding movement usually down a slope between adjacent plies of felt along the bitumen film separating them. It can also take place between gravel surfacing and roofing, between roofing membrane and the insulation, or

A Glossary of Roofing Terms

between the insulation and the roof deck.

Slip Sheet: Sheet material placed between two layers of a roofing system to assure that there is no adhesion between them.

Slope: The incline of a roof surface in degrees, as a slope ratio of fall to run, or as a percentage of fall to run.

Soffit: The underside of any subordinate member of a building. For roofs the underside of a roof overhang.

Softening Point: A measure of the temperature sensitivity of bitumen, by an empirical test that gives the temperature at which a steel ball of specific size falls a definite distance through a disk of the bitumen when the test assembly is heated at a specific rate.

Splitting: The formation of long cracks usually completely through a built-up roofing membrane representing a tension failure of the membrane.

Spray Pond: Intentional ponded water on a roof with a system of piping and jets to spray water above the roof to achieve good evaporative cooling.

Spud: See scraper.

Stabilizer: See filler.

Stack: A vertical vent pipe penetrating above a roof such as that used to provide an escape for foul gases from plumbing fixtures.

Stack Effect: Air flow into a building at the lower levels and out at the higher levels caused by the pressure difference that exists because of the temperature differences of the air masses inside and outside of buildings similar to the phenomenon that produces draft in a chimney.

Stack Venting: The practice of providing small vertical pipe outlets through a roofing membrane to relieve the pressure of water vapour entrapped in the system and with the hope of drying materials such as insulation below the membrane.

Stapling: The use of a specially-designed staple gun to drive staples in place of nails for anchoring roofing materials to nailable decks or other elements of the roofing system.

Starter Strip: Partial-width strip of felt applied at the eaves or other starting line of built-up roofing to serve as the base for the first full course of roofing.

Steep Asphalt: Asphalt of high melting point suitable for steeply-sloped roofs with inclines greater than 1:6. Type 3 as defined by the Canadian Standards Association and Type IV as defined by the American Society for Testing and Materials.

Strainer: A wire, plastic or cast-metal cage placed over the top of a roof drain to prevent debris and leaves on the roof from entering the drain.

Stripping: 1. Narrow widths of felt used to complete flashing details, particularly to cover the edges of metal flanges. 2. The technique of completing flashing details with narrow strips of felt or fabric and hot or cold-applied bitumen.

Substrate: The underlying surface of a roofing membrane. The surface of the deck or the insulation that is the supporting base for the roofing.

Sump: A reservoir forming part of a roof drain. A depression in the roof deck or insulation around a roof drain to provide a water reservoir.

Surfacing: Any aggregate or granular material or coating used as a protective covering on the weather surface of roofing. The protective and traffic-bearing layer of a roof terrace also called the top cover.

System: An assembly of interacting components. A roof system is designed to weatherproof and normally also to insulate the top of a building.

Tanker: A tank truck specially designed with heating and pumping equipment for conveying and dispensing liquid bitumen.

Tar: Black or dark brown liquid or semi-liquid condensates derived from the heating or baking, sometimes called destructive distillation, of wood, peat, oil shale, bone, petroleum, coal or other organic materials. The word is incorrectly used to describe coal tar pitch as in the expression "tar-and-gravel roofing."

Thermal: Relating to heat.

Thermal Bridge: A heat-conductive element in a roof or wall that extends from the warm to the cold side and provides less heat-flow resistance than the adjacent construction. May be of considerable consequences when it passes through the insulation of a well-insulated wall or roof.

Thermal Conductance: A unit of heat flow and a measure of the heat-insulating efficiency of a material or component of a particular thickness. The symbol "C" and units $W/(m^2 \cdot °C)$ are used.

Thermal Conductivity: The basic unit of heat flow, being the amount of heat energy conducted through a unit area of unit thickness in unit time with unit temperature difference between the faces. Expressed in watts (Joules per second) per square metre per metre thickness per degree Celsius temperature difference. The symbol is a small "k" referred to as k-value or k-factor and the units reduce to $W/(m \cdot °C)$. $C = k/n$ where n is thickness in metres.

Thermal Resistance: A Measure of the resistance to heat flow of a material or component of construction of a particular thickness. The symbol "R" is used but is also used for the total resistance of a number of components of materials combined in a roof system. $R = 1/C$ for a particular material and the units are square metres degree Celsius per watt $(m^2 \cdot °C)/W$.

Thermal Resistivity: The basic property of a material's resistance to heat flow through a unit area of unit thickness for unit temperature difference between the faces. It is $1/k$ and the units are metre degree Celsius per watt $(m \cdot °C)/W$.

Thermal Transmittance: A measure of the heat conducted through a unit of a roof system (or other building element) in unit time with unit temperature difference between inside and outside. Also called the coefficient of heat transfer or "U" value for the roof. It is the time rate of heat flow under assumed steady-state conditions that enables one to calculate the heat loss and temperature conditions at any point in a roof system for some particular conditions. Its units are watts per square metre degree Celsius $W/(m^2 \cdot °C)$. It is the reciprocal of the total resistance for the system obtained by adding all the individual resistances of the components of the system including surface and air space resistances. $U = 1/R$ (total).

Thermal Shock: A stress-producing phenomenon thought to result from sudden temperature changes to a roof membrane that takes place with rapid weather changes.

Tin Caps: Small flat metal discs used with nails for securing roofing felts to nailable decks.

Trimmer: The beam or roof joist into which a header is framed in the formation of a roof opening.

Truss: A combination of members such as beams, bars and ties, usually arranged in triangular units, to form a rigid framework for supporting loads over relatively long spans as in wide span roof construction.

Truncated: A hip type of roof terminating in a flat roof.

Underlay: A material, usually felt, used in covering a roof deck before the roofing materials are supplied.

Valley: 1. The horizontal line formed along the depressed angle at the bottom of two inclined roof surfaces. 2. The sloping line of the depressed interior angle formed by two inclined roofs whose eaves meet at right angles.

Vapour: A substance in gaseous state. In relation to building it generally refers to water vapour.

Vapour Barrier: A material, usually in sheet form, used to retard the passage of water vapour into a wall or roof.

Vapour Migration: The movement of water molecules from a region of high to one of lower vapour pressure through the walls and roofs of buildings.

Vapour Permeability: The rate at which water vapour will diffuse or permeate through a unit area in unit time with unit vapour pressure difference across a unit thickness of a material. The units are nanograms per square metre per metre of thickness per second of time per pascal of pressure difference. The symbol is $\bar{\mu}$ and the units are written ng/(Pa•s•m).

Vapour Permeance: The rate at which water vapour will diffuse through a material of a particular thickness. The symbol is M and the units nanograms per square metre per second per pascal vapour pressure difference written ng/(Pa•s•m^2). $M = \bar{\mu}/l$ where l = thickness in metres.

Vapour Resistance: A measure of the resistance to water-vapour flow. Vapour resistance is the reciprocal of permeance = $1/M$ and the units are written (Pa•s•m^2)/ng.

Vermiculite: An aggregate used for lightweight insulating concrete and roof fills, formed by the expansion of mica rock through heating.

Viscosity: The internal resistance offered by a fluid to change of shape or to relative motion or flow of its parts. The flow characteristics of bitumen measured in centistokes. Asphalt may vary from 30 to 500 centistokes when heated from 175°C to 260°C depending on the asphalt type.

Waterproofing: A material used to treat or cover a building element or component to prevent permeation of water. The act of making something impervious to water.

Wrinkling: Small ridges formed at the surface of roofing membranes similar to ridging.

24

Metric Conversion Tables

Note: For an explanation of the International System of Units (SI) refer to NBS Special Publication 330-1974 edition or later, U.S. Department of Commerce/National Bureau of Standards.

The Metric Practice Guide prepared by Canadian Standards Association, published 1973, explains the use of metric units (CSA Z234.1-1973), Canadian Standards Association, 178 Rexdale Boulevard, Rexdale, Ontario Canada M9W 1R3.

Length

inches	in	(×)	2.54	=	centimeters	cm
feet	ft	(×)	30.48	=	centimeters	cm
yards	yd	(×)	0.91	=	meters	m
miles	mi	(×)	1.60	=	kilometers	km
millimeters	mm	(×)	0.04	=	inches	in
centimeters	cm	(×)	0.39	=	inches	in
meters	m	(×)	3.28	=	feet	ft
kilometers	km	(×)	0.62	=	miles	mi

Metric Conversion Tables

Area

square inches	in^2	(×)	6.45	=	square centimeters	cm^2
square feet	ft^2	(×)	0.09	=	square meters	m^2
square yards	yd^2	(×)	0.83	=	square meters	m^2
square miles	mi^2	(×)	2.58	=	square kilometers	km^2
acres		(×)	0.4	=	hectares	ha

square centimeters	cm^2	(×)	0.16	=	square inches	in^2
square meters	m^2	(×)	10.76	=	square feet	ft^2
square meters	m^2	(×)	1.19	=	square yards	yd^2
square kilometers	km^2	(×)	0.38	=	square miles	mi^2
hectares	ha	(×)	2.47	=	acres	

Mass

ounces	oz	(×)	2.34	=	grams	g
pounds	lb	(×)	0.45	=	kilograms	kg
short tons	(2000 lb)	(×)	0.90	=	tonnes	t

grams	g	(×)	0.035	=	ounces	oz
kilograms	kg	(×)	2.20	=	pounds	lb
tonnes	t	(×)	1.10	=	short tons	

Volume U.S.

fluid ounces	(×)	29.57	=	milliliters	ml
pints	(×)	0.47	=	liters	l
quarts	(×)	0.95	=	liters	l
gallons	(×)	3.78	=	liters	l

Volume Imperial

fluid ounces	(×)	28	=	milliliters	ml
pints	(×)	0.57	=	liters	l
quarts	(×)	1.14	=	liters	l
gallons	(×)	4.54	=	liters	l

Volume

cubic feet	ft^3	(×)	0.03	=	cubic meters	m^3
cubic yards	yd^3	(×)	0.76	=	cubic meters	m^3
cubic meter	m^3	(×)	35.315	=	cubic feet	ft^3

Temperature (exact)

Fahrenheit	°F	to Celsius	°C	—	(°F − 32) × 5/9
Celsius		to Fahrenheit		—	9/5 + 32

Temperature (approximate)

Celsius × 2 plus 30 approx. = Fahrenheit

Force

newton (N) = 0.225 lbf
pound-force (lbf) = 4.448 N

Metric Conversion Tables

Pressure, stress

newton per square metre (N/m^2) = 1.4504×10^{-4} lbf/in^2
pound-force per square inch (lbf/in^2) = 6894.8 N/m^2

Density (mass/volume)

kilogram per cubic metre (kg/m^3)	=	0.062 lb/ft^3
pound per cubic foot (lb/ft^3)	=	16.019 kg/m^3
gram per cubic centimetre (g/cm^3)	=	0.036 lb/in^3
pound per cubic inch (lb/in^3)	=	27.680 g/cm^3

Heat

Specific heat	Btu/lb °F	(×)	4.184	=	kJ/kg °C
Thermal conductivity	Btu/hr ft °F	(×)	1.730	=	W/m °C
Heat conductance	Btu/hr ft °F	(×)	5.674	=	W/m °C
Heat flow	Btu/hr ft^2	(×)	3.152	=	W/m^2
Heat transfer	Btu/ft^2	(×)	11.348	=	kJ/m^2

Index

Air leakage, 14–15
Air/vapor barriers, 89
American Society of Heating,
    Refrigeration, and Air
    Conditioning Engineers, 14
Architectural designs:
    effect on roofing, 4–6
Art of building, 12
Asphalt:
    behavior characteristics, 63
    emulsion, 108, 155, 166, 192,
        193, 194
    equiviscous temperature, 64
    properties of, 64, 65
    quantities in roofs, 106, 107,
        109
    selection, 105, 106
    softening points and roof inclines,
        60, 63, *Table 6.1, 64*
    water absorption, 7

Back nailing, 105
Baker, Maxwell C., 85, 86, 132
Barrett specifications:
    roof inclines, *Table 1.1, 8*
Bitumen heating, 9
Blisters, 9, 34, 43, 91, 100, 101,
    153, 158, 164, 166, 168,
    169

Cap sheets (mineral surfaced), *Fig.
    14.4, 157*
Cloquet, Minnesota, 5
Cold built-up roofing:
    decks and substrates, 193, 194
    equipment required, 194, 195
    history, 191, 192
    product description, 192, 193
    specification and application, 194
Concrete:
    roof decks, 43

Index

Concrete *(cont.)*
    thermal movement, 46
    water content, 43
Cooling, 76
Critical path construction scheduling, 9
Cut-out samples, *Fig. 14.5, 158*

Decks for roofs:
    types, 24, 25, 108, 109
Dew point temperatures, *Table 2.2, 16*
Drainage systems:
    design of, 125, 126
    general information, 125, 126
    installation, *Fig. 12.1, 127*

Exterior environment, 17
    hail, 20
    sunshine, 17
    water, 18, 19
    wind, 18

Factory Mutual Engineering Corporation, 198
Failures (building and roofing), 3, 4, 93, 94, 97
Federal Construction Council Report (1955), 11
Fire resistance and ratings, 197, 198
Flashings:
    basic rules, 129–131
    thermal expansion, *Fig. 13.1, 132*
Flashing details:
    brick and block walls, *Fig. 13.10, 141*
    cant strip, *Fig. 13.5, 136, Fig. 13.6, 137*
    expansion joints, *Fig. 13.11, 142*
    gravel stops, *Fig. 13.3, 134*
    gravel stops and gutters, *Fig. 13.4, 135*
    joints, *Fig. 13.2, 133*
    mechanical equipment stand, *Fig. 13.12, 143*
    miscellaneous, *Figs. 13.13 to 13.18, 144–149*
    poured concrete wall, *Fig. 13.9, 140*
    scupper, *Fig. 13.7, 138*
    wall flashing (caulked), *Fig. 13.8, 139*
Fricklas, Richard L., 12

Glossary, 209
Gravel surfacing, 108, 155, 156, 160
Guarantees, 203–206
    example, 204–206

Insulation:
    Baker, Maxwell C., effect of insulation on roofing, 85, 86
    batts, blankets, loose fill (advantages), 71
    in Conventional systems, 76, 77
    danger to roof systems, 74
    early varieties, 9
    example of failure, 78
    frank statement, 86
    function of, 16, 85, 86
    hazards, 74, 80, 81, *Fig. 7.1, 82*
    heat transfer, 72, *Table 7.1, 73*
    kinds used, 71, 73, 74
    location in buildings, 71
    moisture detection, 77, 78
    need for, 75, 76
    physical properties, 72
    on steel decks, 38–41
    thickness (maximum), 74
    urethane foam, 171, 173
    wetting, 118
Interior environment, 13

Jones, P. M., 7

Life expectancy, 203–206
Literature, 207, 208

Maintenance of roofs, 199–201
Materials for BUR (history), 6
    handling and storage, 159–161
Mill construction, 8
Moisture detection, 77, 78
    in insulation, 77–80
    meters, 77

Mopping, 25, 152, 154, 155, 156
Mosely, G. Norman, 83
Moss growth, 61

National Roofer (1955), 11
National Roofing Contractors Association, 11, 86

Performance of old roofs, 9
pH, definition of, 60
Pitch (coal-tar):
    advertising and propaganda, 7
    properties, 6, 65, 66, 105
    selection for roofs, 107
    self healing, 8
    quantities in roofs, 107
    water absorption, 7, 8
Ponded water:
    consequences, 18–20
    examples, *Fig. 5.1*, *57*, *58*
Portland Cement Association, 52
Portneuf, Quebec, 5
Propaganda, 7

Reflective surfacing, *Table 7.3*, *87*, 76
Relative humidity, definition, 14
Re-roofing:
    deck construction, 167
    examples, 167-171
    —apartment buildings, 169, 170
    —building materials warehouses, 168
    —factory building (no. 7), 170, 171
    —office building, 168
    —paper mill storage warehouse, 168
    —plywood mill, 168
    guidelines, 164, 165
    removal of old roof, 163
    resurfacing, 164, 166
Richardson, H. H., 4
Roll roofing:
    mineral surfaced, 70
    selvage edge, 70
    smooth surfaced, 70

Roof decks:
    application of roofing on, *Table 4.1*, *26*, *27*
    asbestos cement cavity, structural details, *Fig. 4.18*, *53*, *Fig. 4.19*, *54*
    boards, 25, 28, 29
    desirable qualities, 23, 24
    lightweight concretes, 51, 52
    old roof decks, 8
    plywood over T&G decking, 31, 33
    plywood on wood joists, 29, 30, 31, 34
    poured concrete, 43
    —advantages, 44, 46
    —important considerations, 46, 47
    poured gypsum, 41
    —roof specification over, *Fig. 4.8*, *42*
    precast concrete:
    —deficiencies, 47–49
    —details of long span slabs, *Fig. 4.14*, *49*, *Fig. 4.15*, *50*
    —details of short span slabs, *Fig. 4.11*, *47*, *Fig. 4.12*, *48*
    —important considerations, 50
    precast gypsum, 43
    —architectural details, *Fig. 4.10*, *45*
    —roof specification over, *Fig. 4.9A*, *44*
    steel:
    —important considerations, 37, 38, 40, 41
    —insulation fasteners, *Table 4.2*, *37*, *Fig. 4.5*, *39*, *40*
    —problems with, 35–37
    —insulation on, 37, *Fig. 4.4*, *38*, *Fig. 4.5*, *39*, *40*
    types, 24, 25
    ventilation of (frame construction), *Fig. 4.2B*, *30*, *Fig. 4.2C*, *31*
    wood fiber and cement slabs, 52
Roof inclines:
    acceptable minimum for water flow, 55
    Barrett specifications, *Table 1.1*, *8*

Index

Roof inclines *(cont.)*
  danger from chemicals, 60, 61
  dead level roofs, 61, 62
  general rules, 57
  minimum for smooth surfaced roof specs., 56
  moss growth, 61
  problems with level decks, 55–57
  steep roofs, 60

Roof membranes:
  cold built-up roofing, 191, 195
  single ply, 175–189
    —CGSB standards, *Table 17.6, 188*
    —contaminants, *Table 17.5, 187*
    —generic names, *Table 17.1, 177*
    —insulation approved by manufacturers, *Table 17.4, 186*
    —introduction, 175, 176
    —manufacturers, 176, *Table 17.1, 177*
    —partial product specs., *Table 17.3, 179–185*
    —product names, *Table 17.2, 178*
    —references, 189

Roof systems:
  application procedures.
    —flood coat and gravel, 155, *Fig. 14.3, 156*
    —heating bitumen, 151, 152
    —laying base sheets, 153, 154
    —laying felt, *Fig. 14.1, 152,* 153
    —mopped surface coats, 155
    —stripping felts, 154, 155
    —vapor barriers and insulation, 155, 157
  asphalt quantities, 106, 107
  asphalt selection, 105, 106
  back nailing felt, 105
  base sheet application, *Fig. 10.2, 103*
  base sheet, use of, 101, 102
  decks, 24, 25, 108, 109
  development of, 100
  failure rate, 93
  maintenance, 199–201
  pitch quantities, 107
  pitch selection, 107
  priming decks, 100, 101
  protected membrane roofs, 110
    —appropriate applications, 112
    —illustrations, *Fig. 10.4, 113, 114, Fig. 10.5, 115, Fig. 10.6, 116*
    —temperature curves, 110, 112, 115, *Fig. 10.7, 117,* 118
  research, 94–96
  roofing felt and ply sheets, 102, 104, 105
  state of the art (1979–1982), 93
  surfacing materials, 107, 108
    —flood coat and gravel, 155
    —gravel application, *Fig. 14.3, 156*
    —gravel, *Fig. 10.3, 108*
    —handling and application, 160
    —mopped coats, 155, *Fig. 14.3, 156*
    —reflectivity, *Table 7.3,* 87
    —thermal expansion, 81, 104
    —typical roof membranes, 109, 110
    —wood deck specification (old), 101

Roof traffic, 58, 59
  examples, 59

Roofing bonds, 10, 203, 204

Roofing felt:
  asbestos, asphalt and coal-tar saturated, 68, 69
  —supply of, *fn. 1,* 68
  glass fiber:
  —ASTM designation, 69
  —CSA standard, 69, 70
  organic (asphalt saturated), 66

Roofing felt:
  coefficients of thermal expansion, *Table 7.2,* 81, *Table 10.2, 104*
  damage to, 159, *Fig. 15.1,* 160
  direction over insulation, *Fig. 7.1,* 82
  direction over plywood, *Fig. 4.2,* 29, *Fig. 4.3,* 33
  laying patterns, 83

Index

lining, *Fig. 6.1*, 67, 68
shingle mopping, *Fig. 7.2*, 84
tar saturated, 28, 66, 68
Roofing Industry Educational Institute, 12
Roofing problems:
  contributors to, 96, 97
  nineteen fifty-five (1955), 11
  reasons for, 97–100
Roofing:
  splits, 83, 85
  —references, 88
  steel and aluminum:
  —advantages, 119
  —decks for, 120
  —forming of, *Figs. 11.1 to 11.5*, *122–124*
  —insulation of, 120
  —vapor barriers, 120
  self healing, 8
  shrinkage of felts, 81, 83, 85
  single ply membranes, 175–189
  solar radiation, 17
  standards:
  —bonds (roofing), 10
  —materials, 10

Sullivan, Louis Henry, 5
Surface temperatures, 16, 87, 106

Tables:
  application details for various roof decks, *Table 4.1*, *26, 27*
  Asphalt types, *Table 6.1*, *64*
  coefficient of solar reflectivity, *Table 7.3*, *87*
  coefficients of thermal expansion, *Table 7.2*, *81*, *Table 10.2*, *104*
  dew point, *Table 2.2*, *16*
  felt laps and exposures, *Table 10.1*, *104*
  ASTM designation and CSA standards for *Glass* mat for roofing, 69
  fasteners for insulation on steel decks, *Table 4.2*, *37*
  insulation—amount required, *Table 7.1*, *73*
  roof inclines, *Table 1.1*, *8*
  roof inclines for asphalt, *Table 10.3*, *105*
  roof inclines for pitch, *Table 10.4*, *105*
  single ply roofing, *Tables 17.1 to 17.6*, *177–188*
  vapor pressure, *Table 2.1*, *15*
Typical roof membranes, 109, 110
Temperatures:
  roof surfaces, 16, 76, 87
  in roof systems, 117
Thermal expansion of metals, *Fig. 13.1*, *132*

Underwriters Laboratories Inc., 41, 198
Urethane foam, 171, 173

Van der Rohe, Mies, 5
Vapor barriers and insulation, 155, 157
Vapor pressure, definition, 14
  vapor pressures, *Table 2.1*, *15*
Ventilation:
  roof decks, frame construction, *Fig. 4.2, A, B, C*, *29–31*
  roof systems, 91, 92
Vermiculite/asphalt mix, 9

Water:
  cooling, 19, 76
  effect on flat roofs, 19
  in insulation, 77–80
  in roofing materials, 159
Weather data sources, 20, 21
Wind, 18
Wood decks (boards), 25, *Fig. 4.1*, *28*